Reliability-Based Mechanical Design
Volume 2

Component under Cyclic Load and Dimension
Design with Required Reliability

Synthesis Lectures on Mechanical Engineering

Synthesis Lectures on Mechanical Engineering series publishes 60–150 page publications pertaining to this diverse discipline of mechanical engineering. The series presents Lectures written for an audience of researchers, industry engineers, undergraduate and graduate students.

Additional Synthesis series will be developed covering key areas within mechanical engineering.

Reliability-Based Mechanical Design, Volume 2: Component under Cyclic Load and Dimension Design with Required Reliability
Xiaobin Le
2019

Reliability-Based Mechanical Design, Volume 1: Component under Static Load
Xiaobin Le
2019

Solving Practical Engineering Mechanics Problems: Advanced Kinetics
Sayavur I. Bakhtiyarov
2019

Natural Corrosion Inhibitors
Shima Ghanavati Nasab, Mehdi Javaheran Yazd, Abolfazl Semnani, Homa Kahkesh, Navid Rabiee, Mohammad Rabiee, Mojtaba Bagherzadeh
2019

Fractional Calculus with its Applications in Engineering and Technology
Yi Yang and Haiyan Henry Zhang
2019

Essential Engineering Thermodynamics: A Student's Guide
Yumin Zhang
2018

Engineering Dynamics
Cho W.S. To
2018

Solving Practical Engineering Problems in Engineering Mechanics: Dynamics
Sayavur I. Bakhtiyarov
2018

Solving Practical Engineering Mechanics Problems: Kinematics
Sayavur I. Bakhtiyarov
2018

C Programming and Numerical Analysis: An Introduction
Seiichi Nomura
2018

Mathematical Magnetohydrodynamics
Nikolas Xiros
2018

Design Engineering Journey
Ramana M. Pidaparti
2018

Introduction to Kinematics and Dynamics of Machinery
Cho W. S. To
2017

Microcontroller Education: Do it Yourself, Reinvent the Wheel, Code to Learn
Dimosthenis E. Bolanakis
2017

Solving Practical Engineering Mechanics Problems: Statics
Sayavur I. Bakhtiyarov
2017

Unmanned Aircraft Design: A Review of Fundamentals
Mohammad Sadraey
2017

Introduction to Refrigeration and Air Conditioning Systems: Theory and Applications
Allan Kirkpatrick
2017

Resistance Spot Welding: Fundamentals and Applications for the Automotive Industry
Menachem Kimchi and David H. Phillips
2017

MEMS Barometers Toward Vertical Position Detecton: Background Theory, System Prototyping, and Measurement Analysis
Dimosthenis E. Bolanakis
2017

Engineering Finite Element Analysis
Ramana M. Pidaparti
2017

Reliability-Based Mechanical Design, Volume 2: Component under Cyclic Load and
Dimension Design with Required Reliability

Xiaobin Le

ISBN: 978-3-031-79640-1   paperback
ISBN: 978-3-031-79641-8   ebook
ISBN: 978-3-031-79642-5   hardcover

DOI 10.1007/978-3-031-79641-8

A Publication in the Springer series
SYNTHESIS LECTURES ON MECHANICAL ENGINEERING

Lecture #21
Series ISSN
Print 2573-3168   Electronic 2573-3176

Reliability-Based Mechanical Design Volume 2

Component under Cyclic Load and Dimension Design with Required Reliability

Xiaobin Le
Wentworth Institute of Technology

SYNTHESIS LECTURES ON MECHANICAL ENGINEERING #21

ABSTRACT

A component will not be reliable unless it is designed with required reliability. *Reliability-Based Mechanical Design* uses the reliability to link all design parameters of a component together to form a limit state function for mechanical design. This design methodology uses the reliability to replace the factor of safety as a measure of the safe status of a component. The goal of this methodology is to design a mechanical component with required reliability and at the same time, quantitatively indicates the failure percentage of the component. *Reliability-Based Mechanical Design* consists of two separate books: *Volume 1: Component under Static Load*, and *Volume 2: Component under Cyclic Load and Dimension Design with Required Reliability*.

This book is *Reliability-Based Mechanical Design, Volume 2: Component under Cyclic Load and Dimension Design with Required Reliability*. It begins with a systematic description of a cyclic load. Then, the books use two probabilistic fatigue theories to establish the limit state function of a component under cyclic load, and further to present how to calculate the reliability of a component under a cyclic loading spectrum. Finally, the book presents how to conduct dimension design of typical components such as bar, pin, shaft, beam under static load, or cyclic loading spectrum with required reliability. Now, the designed component will be reliable because it has been designed with the required reliability.

The book presents many examples for each topic and provides a wide selection of exercise problems at the end of each chapter. This book is written as a textbook for senior mechanical engineering students after they study the course Design of Machine Elements or a similar course. This book is also a good reference for design engineers and presents design methods in such sufficient detail that those methods are readily used in the design.

KEYWORDS

reliability, reliability-based design, mechanical component, mechanical design, computational method, numerical simulation, static load, cyclic load, fatigue damage, limit state function, failure, safety, probability, the P-S-N curve approach, the K-D probabilistic fatigue damage model

To my lovely wife, Suyan Zou,
and to my wonderful sons, Zelong and Linglong

Contents

Preface

Reliability-Based Mechanical Design consists of two separate books: *Volume 1: Component under Static Load* and *Volume 2: Component under Cyclic Load and Dimension Design with Required Reliability*.

Volume 1 consists of four chapters and Appendix A. They are:

- Chapter 1: Introduction to Reliability in Mechanical Design;

- Chapter 2: Fundamental Reliability Mathematics;

- Chapter 3: Computational Methods for the Reliability of a Component;

- Chapter 4: Reliability of a Component under Static load; and

- Appendix A: Samples of MATLAB Programs.

Volume 2 consists of three chapters and two appendixes. They are:

- Chapter 1: Introduction and Cyclic Loading Spectrum;

- Chapter 2: Reliability of a Component under Cyclic Load;

- Chapter 3: The Dimension of a Component with Required Reliability;

- Appendix A: Three Computational Methods for the Reliability of a Component; and

- Appendix B: Samples of MATLAB Programs.

The first book discusses fundamental concepts for implementing reliability in mechanical design and the reliability of a component under static load. The second book presents more advanced topics, including the reliability of a component under cyclic load and the dimension design with required reliability.

This is *Reliability-based Mechanical Design, Volume 2: Component under Cyclic Load and Dimension Design with Required Reliability*. It is recommended that Volume 1 should be read first before Volume 2 because it provides fundamental concepts and computational methods for implementing reliability in mechanical design and the reliability of a component under static load.

This book presents how to determine reliability, and quantitively predict the failure percentage of a component under cyclic load. This book also presents how to design component dimension with required reliability for a component under static load or cyclic load. Therefore, the component will be reliable during its service under the specified load because it has been designed with required reliability.

This book is based on the author's recent research and a series of lecture notes of an elective course for senior mechanical students. This book is written as a textbook for senior mechanical students. Every topic is discussed in sufficient detail and demonstrated by many examples so that students or design engineers can readily use them in mechanical design. At the end of each chapter, there is a wide selection of exercise. This book can also be used for a graduate student course and a reference book for design engineers.

This book consists of three chapters and two appendixes. A concise summary of each chapter are as follows.

- Chapter 1: Introduction and Cyclic Load Spectrum

  This chapter presents a systematic description of a cyclic load. Six models of cyclic loading spectrum will be presented and can be used to describe any type of cyclic load.

- Chapter 2: Reliability of a Component under a Cyclic Load

  This chapter describes how to establish the limit state function of a component under a cyclic load, and then how to determine the reliability of a component under such cyclic load. The book presents two fatigue theories to calculate the reliability of a component under cyclic load. The first theory is the P-S-N (Probailitis-Stress-Number of cycles) curve approach. The second theory is the probabilistic fatigue damage model (the K-D model). Five typical component cases under cyclic load presented in this chapter include bar under cyclic axial load, pin under cyclic direct shearing, shaft under cyclic torsion, beam under cyclic bending, and a rotating shaft under cyclic combined loads.

- Chapter 3: The Dimension of a Component with Required Reliability

  This chapter presents how to design the dimension of a component with required reliability under static load or cyclic load. For the dimension of a component under cyclic load, the second fatigue theory, that is, the K-D model is mainly used. Five typical component dimension design with required reliability presented in this chapter include bar under axial load or cyclic axial load, pin under direct shearing or cyclic direct shearing, shaft under torsion or cyclic torsion, beam under bending or cyclic bending, and a component under combined loads or cyclic combined loads.

- Appendix A: Computational Methods for Calculating the Reliability of a Component

  Appendix A concisely describes the procedures of the Hasoder-Lind (H-L) method, the Rachwitz-Fiessler (R-F) method, and the Monte Carlo method for calculating the reliability of a component, which has been presented in details in the first book: *Reliability-based Mechanical Design, Volume 1: Component under Static Load*.

- Appendix B: Samples of six MATLAB Programs

Appendix B provides six MATLAB program as a reference, including three programs for the calculation of reliability and another three programs for dimension design with required reliability.

This book could not have been completed and published without lots of encouragement and help. First, I sincerely thank Mechanical Department Chairman and Professor Mickael Jackson at the Wentworth Institute of Technology, whose encouragement motivated me to open two technical elective courses about the reliability in mechanical engineering. Second, I sincerely thank Professors Anthony William Duva and Richard L. Roberts for reviewing some of the manuscripts. Third, I sincerely thank Morgan & Claypool Publishers and Executive Editor Paul Petralia for helping with this publication. Finally, I sincerely thank my lovely wife, Suyan Zou. Without her support, I could not have completed this book.

Xiaobin Le
October 2019

CHAPTER 1

Introduction and Cyclic Loading Spectrum

1.1 INTRODUCTION

Reliability-Based Mechanical Design consists of two separate books: *Volume 1: Component under Static Load* and *Volume 2: Component under Cyclic Load and Dimension Design with Required Reliability.*

This book is Volume 2. It is recommended that Volume 1 should be read first before Volume 2 since fundamental concepts of probability theory and their implementation in mechanical design, as well as construction of the limit state function of a component under load are discussed in detail. The following are some concise notes on topics previously discussed in Volume 1 but that will be frequently used in this book.

Fundamental reliability mathematics, which discusses the fundamental concepts and definitions of probabilistic theory, is discussed in detail in Chapter 2 of Volume 1 [1]. The purpose of this is to provide a foundation and basic understandings for implementing probability theory in reliability-based mechanical design.

Computational methods of the reliability of a component, which discuss several computational methods including the Hasoder–Lind (H-L) method, the Rachwitz–Fiessler (R-F) method, and the Monte Carlo method, are discussed in Chapter 3 of Volume 1 [1]. The concise description of their procedures and flowcharts are presented in Appendix A of this book.

In reliability-based mechanical design, component geometric dimensions, loads, and stress concentration factor on a component are typically treated as random variables. These are discussed in detail in Chapter 4 of Volume 1 [1]. Since these will be frequently used in this book, we concisely describe them here.

Component geometric dimension is a random variable because of its dimension tolerance. It is typically treated as a normally distributed random variable d. According to the definition of dimension tolerance, the components' dimension inside the dimension tolerance range $(d + t_L, d + t_U)$ will be accepted. For a normal distribution, the probability of event $(\mu_d - 4\sigma_d \leq d \leq \mu_d + 4\sigma_d)$ will be 99.9968%. This event can be used to represent the dimension tolerance range with a very small error (0.0032%). Therefore, the mean and standard deviation of a normally distributed dimension random variable d can be determined per Equation (1.1):

$$
\begin{aligned}
\mu_d &= \frac{(d + t_L) + (d + t_U)}{2} = d + \frac{t_L + t_U}{2} \\
\sigma_d &= \frac{(d + t_U) - (d + t_L)}{8} = \frac{t_U - t_L}{8},
\end{aligned}
\tag{1.1}
$$

where μ_d and σ_d are the mean and standard deviation of a normally distributed dimension d.

The type of distribution and its corresponding distribution parameters of an external load will be calculated per the design specifications of a design case. If a load P such as concentrated force, concentrated moment, or torque, is expressed as a range of value such as (P_{low}, P_{up}), it could be treated as a normally distributed random variable. We can use the same reasoning and similar equation as Equation (1.1) to determine its mean and standard deviation:

$$
\begin{aligned}
\mu_P &= \frac{\left(P_{low} + P_{up}\right)}{2} \\
\sigma_P &= \frac{\left(P_{up} - P_{low}\right)}{8},
\end{aligned}
\tag{1.2}
$$

where μ_P and σ_P are the mean and standard deviation of a normally distributed load P.

The static stress concentration factor is a function of geometric shape and dimension. Since the geometric dimension is a random variable, the stress concentration factor is also a random variable and typically follows a normal distribution. We can use the following equations to determine the mean and the standard deviation of stress concentration factor:

$$
\begin{aligned}
\gamma_K &= 0.05 \\
\mu_K &= K_{Table} \\
\sigma_K &= \gamma_K \times \mu_K = 0.05 K_{Table},
\end{aligned}
\tag{1.3}
$$

where γ_K is the coefficient of variance of the static stress concentration factor. K_{Table} is the stress concentration factor obtained from tables in current design handbooks or design books. μ_K and σ_K are the mean and standard deviation of normally distributed static stress concentration factors.

This book consists of three chapters and two appendixes. The concise outlines of each chapter are as follows.

- Chapter 1: Introduction and Cyclic Loading Spectrum

  This chapter explains the connection of this book with the first book: *Reliability-Based Mechanical Design, Volume1: Component under Static Load*. Then, this chapter presents a systematic description of a cyclic load. Six models of cyclic loading spectrum will be presented and can be used to describe any type of cyclic load.

- Chapter 2: Reliability of a Component under a Cyclic Load

This chapter describes how to establish the limit state function of a component under a cyclic load, and then how to determine the reliability of a component under such cyclic load. The book presents two fatigue theories to calculate the reliability of a component under cyclic load. The first theory is the P-S-N (Probabilidtic-Stress-Number of cycles) curve approach. The second theory is the probabilistic fatigue damage model (the K-D model). Five typical component cases under cyclic load presented in this chapter include bar under cyclic axial load, pin under cyclic direct shearing, shaft under cyclic torsion, beam under cyclic bending, and a rotating shaft under cyclic combined loads.

- Chapter 3: The Dimension of a Component with the Required Reliability

This chapter presents how to design the dimension of a component with required reliability under static load or cyclic load. For the dimension of a component under cyclic load, the second fatigue theory, that is, the K-D model is mainly used. Five typical component dimension design with required reliability presented in this chapter include bar under static axial load or cyclic axial load, pin under static direct shearing or cyclic direct shearing, shaft under static torsion or cyclic torsion, beam under static bending or cyclic bending, and a component under combined static loads or cyclic combined loads.

- Appendix A: Computational Methods for Calculating the Reliability of a Component

This appendix concisely describes the procedure of the H-L, R-F, and Monte Carlo methods for calculating the reliability of a component, which has been presented in detail in Volume 1 [1].

- Appendix B: Samples of Six MATLAB Programs

This appendix provides six MATLAB programs as a reference, including three programs for the calculation of reliability and another three programs for dimension design with required reliability.

1.2   CYCLIC LOADING SPECTRUM

Mechanical devices or systems always have at least one moving component. Due to the repeated functions or stop-start process or mechanical vibration, mechanical components are typically subjected to a cyclic load. A schematic of a cyclic load is depicted in Figure 1.1. The maximum stress $\sigma_{\max}$ and the minimum stress $\sigma_{\min}$ of cyclic stress (loading) are the maximum and minimum values of the cyclic stresses, as shown in Figure 1.1. The mean stress σ_m, the stress amplitude σ_a, and the range of stress σ_r of the cyclic stress (loading) are defined as:

$$\sigma_m = \frac{\sigma_{\max} + \sigma_{\min}}{2} \tag{1.4}$$

$$\sigma_a = \frac{\sigma_{\max} - \sigma_{min}}{2} \tag{1.5}$$

$$\sigma_r = \sigma_{max} - \sigma_{min} = 2\sigma_a. \tag{1.6}$$

The stress ratio S_r is defined as the ratio of the minimum stress σ_{min} to the maximum stress σ_{max}, that is,

$$S_r = \frac{\sigma_{min}}{\sigma_{max}}. \tag{1.7}$$

A cyclic stress curve can be treated as a wave. One complete of the wave such as the minimum point to the adjacent minimum point, or the maximum point to the adjacent maximum point is one cycle of the cyclic stress, as shown in Figure 1.1.

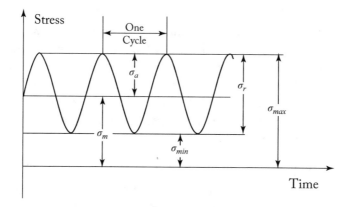

Figure 1.1: A schematic of cyclic stress with a constant stress amplitude.

The magnitude of cyclic stress can be fully defined by any two out of these six variables σ_{max}, σ_{min}, σ_m, σ_a, σ_r, and S_r. The duration of the cyclic stress will be defined by the number of cycles of the cyclic loading. One special cyclic stress that has a zero mean stress is called the fully (completely) reversed cyclic stress, as shown in Figure 1.2. For a fully reversed cyclic stress, it has: $\sigma_{max} = -\sigma_{min}$, and the stress ratio $S_r = -1$. This type of cyclic stress is a special case because lots of fatigue strength data are based on fatigue tests under a fully reversed cyclic stress.

Example 1.1
Cyclic stress has a stress amplitude $\sigma_a = 10$ ksi and the stress ratio $S_r = 0.5$. Calculate the mean stress σ_m, the maximum stresses σ_{max}, the minimum stresses σ_{min}, and the range of stress σ_r of this cyclic stress.

Solution:
Based on Equations (1.5) and (1.7), we have:

$$\sigma_a = 10 = \frac{\sigma_{max} - \sigma_{min}}{2} \tag{a}$$

$$S_r = 0.5 = \frac{\sigma_{min}}{\sigma_{max}}. \tag{b}$$

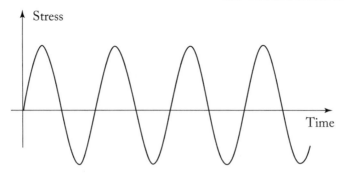

Figure 1.2: A fully reversed cyclic stress.

From Equations (a) and (b), we have:

$$\sigma_{\max} = 40 \text{ (ksi)}, \qquad \sigma_{\min} = 20 \text{ (ksi)}. \tag{c}$$

Based on Equations (1.4) and (1.6) by using information from Equation (c), we have:

$$\sigma_m = \frac{\sigma_{\max} + \sigma_{\min}}{2} = \frac{40 + 20}{2} = 30 \text{ (ksi)}$$

$$\sigma_r = \sigma_{\max} - \sigma_{\min} = 40 - 20 = 20 \text{ (ksi)}.$$

■

There is a lot of different cyclic loads. The cyclic loading spectrum refers to a description of cyclic stress (loading) levels vs. corresponding cycle numbers. Generally, three parameters, including stress amplitude, mean stress, and the number of cycles of the cyclic loading, are used to describe a cyclic loading spectrum fully. However, for fatigue design, a non-zero mean cyclic loading is typically converted into a fully reversed cyclic loading with an equivalent stress amplitude, which will be discussed in detail in Chapter 2. So, two parameters including the fully reversed stress amplitude σ_a and the cycle number n_L of cyclic loading are typically used to describe a cyclic loading spectrum for the reliability calculation of a component under cyclic load. Both stress amplitude σ_a and the cycle number n_L can be a constant (deterministic value) or several constant or random variable. With the reasonable combinations of variations of stress amplitude σ_a and the cycle number n_L, the systematic description of cyclic loading spectrum will include the following six models. These six cyclic loading spectrum models [2] can describe any cyclic loading spectrum.

Model #1: A constant stress amplitude of cyclic loading with a constant cycle number.

Model #1 is the simplest cyclic loading spectrum for fatigue design. For example, the component under design is subjected to a constant cyclic stress amplitude $\sigma_a = 15$ (ksi) with a given cycle number $n_L = 5 \times 10^4$ (cycles).

Model #2: A constant stress amplitude of cyclic loading with a distributed cycle number.

Model #2 is a typical cyclic loading spectrum for a component with a single steady function, that is, a constant cyclic stress amplitude. However, the cycle number n_L of the cyclic loading is treated as a random variable and is described by a probabilistic distribution function. How can this be? It is because, in the reliability-based mechanical design, the components under design are a batch of "identical" components in service. Each component during its service life had one value of the number of cycles. All of those can be used to determine the distribution function of the cycle number. For example, the component under design is subjected to a constant stress amplitude of a fully reversed cyclic stress $\sigma_a = 15$ (ksi) with a normally distributed cycle number n_L, which has a mean $\mu_{n_L} = 3 \times 10^5$ (cycles), and a standard deviation $\sigma_{n_L} = 3500$ (cycles), that is, $n_L = N(3 \times 10^5, 3500)$. Here, the expression $X = N(\mu_x, \sigma_x)$ means that the random variable X is a normal distribution with a mean μ_x, and a standard deviation σ_x.

Model #3: A given fatigue life (cycle number) with a distributed amplitude of a cyclic loading.

Model #3 is a typical cyclic loading spectrum for the component with specified service life, but the fully reversed stress amplitude σ_a varies and can be treated as a random variable, and is described by a probabilistic distribution function. For example, the component under design with a cycle number $n_L = 8 \times 10^4$ (cycles) is subjected to a fully reversed cyclic stress. The stress amplitude σ_a of this cyclic loading follows the uniform distribution between 25 (ksi) and 35 (ksi).

Model #4: Multiple constant amplitudes of cyclic loadings with multiple constant cycle numbers.

Typically, model #4 could be used to describe the cyclic loading spectrum of a machine with several distinguished functions or actions. For example, the cyclic loading spectrum of the component under design is:

$$\text{Cyclic stress level } 1: \sigma_{a1} = 20 \text{ (ksi)}, \; n_{L1} = 260{,}000 \text{ (cycles)}$$
$$\text{Cyclic stress level } 2: \sigma_{a2} = 30 \text{ (ksi)}, \; n_{L2} = 50{,}000 \text{ (cycles)}.$$

Model #5: Multiple constant stress amplitudes of cyclic loadings with multiple distributed cycle numbers.

Model #5 is a common cyclic loading spectrum and can be used to describe many loading conditions for machines with several distinguished functions. For example, the cyclic loading spectrum of the component under design is:

$$\text{Cyclic stress level } 1: \sigma_{a1} = 20 \text{ (ksi)}; \; n_{L1} = N(260{,}000, 10{,}000)$$
$$\text{Cyclic stress level } 2: \sigma_{a2} = 25 \text{ (ksi)}; \; \ln(n_{L2}) = N(8.425, 0.136).$$

Model #6: Multiple distributed stress amplitudes of cyclic loading levels with multiple given cycle numbers.

Model #6 is also a common cyclic loading spectrum and can be used to describe many loading conditions for the machines, the service of which are pre-scheduled. For example, the cyclic loading spectrum of the component under design is:

$$\text{Cyclic stress level } 1 : n_{L1} = 260{,}000 \text{ (cycles)}; \ \sigma_{a1} = N\,(20{,}000, 1890)$$
$$\text{Cyclic stress level } 2 : n_{L2} = 5000 \text{ (cycles)}; \ \ln(\sigma_{a2}) = N(3.25, 0.108).$$

Any cyclic loading for fatigue design can be described by one of the above six fatigue cyclic loading spectrum models. Thus, they are a systematic description of cyclic loading spectrum.

1.3   REFERENCES

[1] Le, Xiaobin, *Reliability-Based Mechanical Design, Volume 1: Component under Static Load*, Morgan & Claypool Publishers, San Rafael, CA, 2019. 1, 3

[2] Le, Xiaobin, The reliability calculation of components under any cyclic fatigue loading spectrum, *ASME International Mechanical Engineering Congress and Exposition, IMECE–70084*, Tampa, FL, November 3–9, 2017. DOI: 10.1115/imece2017-70084. 5

1.4   EXERCISES

1.1. Cyclic stress has a maximum stress $\sigma_{\max} = 60.25$ (ksi) and a minimum stress $\sigma_{\min} = -9.32$ (ksi). Calculate its mean stress, stress amplitude, and stress ratio.

1.2. Cyclic stress has a constant mean $\sigma_m = 15.72$ (ksi) and a stress amplitude $\sigma_a = 25.39$ (ksi). Calculate its maximum stress and the minimum stress of this cyclic stress.

1.3. What is cyclic loading spectrum? Describe one example.

1.4. What causes cyclic stress? Use two examples to explain the lists.

1.5. Describe an example in which model #1 cyclic loading spectrum can be used to describe its cyclic stress.

1.6. Describe an example in which model #2 cyclic loading spectrum can be used to describe its cyclic stress.

1.7. Describe an example in which model #3 cyclic loading spectrum can be used to describe its cyclic stress.

1.8. Describe an example in which model #4 cyclic loading spectrum can be used to describe its cyclic stress.

1.9. Describe an example in which model #5 cyclic loading spectrum can be used to describe its cyclic stress.

1.10. Describe an example in which model #6 cyclic loading spectrum can be used to describe its cyclic stress.

C H A P T E R   2

Reliability of a Component under Cyclic Load

2.1   INTRODUCTION

Metal components under repeated loadings, that is, cyclic load, can fracture even though the component's maximum nominal stress is far less than ultimate material strength or yield strength. This type of failure is fatigue failure. In industries, more than 90% of metal component failure is due to fatigue failure.

A fatigue issue with the number of cycles at failure between 1 and 10^3 is generally classified as low-cycle fatigue. When the number of cycles at failure is more than 10^3, it is high-cycle fatigue. This chapter will only focus on high-cycle fatigue, which is the typical case for mechanical component design in industries.

The component design, that is, determination of component geometric dimension under cyclic loading with the specified reliability will be discussed in Chapter 3. This chapter will show and explain how to calculate the reliability of a component under cyclic loading. The reliability calculation of a component with an infinite life will be discussed in Section 2.7. Two different probabilistic fatigue theories will be used to conduct the reliability calculation of a component with a finite fatigue life. One theory is the P-S-N curves approach, which will be discussed and explained in Section 2.8. Another theory is the probabilistic fatigue damage theory, which includes fatigue strength index K and the fatigue damage index D. This probabilistic fatigue damage theory can be called as the K-D model, which will be discussed and explained in Section 2.9.

This chapter will present and discuss different methods to determine the reliability of a component under any cyclic load with plenty of examples.

2.2   FATIGUE DAMAGE MECHANISM

Fatigue phenomena were first discovered and studied during the 19th century with the arrival of machines and freight vehicles during the industrial revolution [1]. Fatigue is defined as "failure under a repeated or varying loading, which never reaches a level sufficient to cause failure in a single application."

Fatigue failure of a metal component under cyclic loading is a complicated phenomenon, and only partially understood [2]. However, we have a fundamental understanding of fatigue failure or fatigue damage. Fatigue damage is the weakening of a metal material due to a gradually crack propagation of inherent existing microscopic cracks or defect inside or on the surface of the metal component under repeated cyclic loading. Without a crack inside or on the surface of a metal component, there is no fatigue. If the magnitude of cyclic loading is not big enough to generate a crack propagation, there will be no fatigue.

The fatigue damage mechanism can be typically described by the following four stages. Let use a microscopic crack on the surface to explain and demonstrate these. Figure 2.1 shows a magnified microscopic crack on the surface of a component under a fully reversed cyclic bending moment. In this example, let us assume that the nominal normal stress due to the bending stress is 20 ksi, and the material yield strength is 60 ksi.

1. Crack initiation. There are always lots of randomly distributed defects inside a component such as voids and dislocations and on the surface of a component such as manufacturing scratches [3]. A fatigue crack will typically initiate at a microscopic crack or defect inside or on the surface of a component. As shown in Figure 2.1a, the bending moment tries to open the microscopic crack "A", which will have very high stress because of the sharp tip of the microscopic crack "A". Let us assume that the stress concentration factor in this situation is 3.5. So, the maximum stress at the tip of the microscopic crack "A" could be $3.5 \times 20 = 70$ ksi and will be larger than the material yield strength. The material at the tip of the crack "A" will start yielding and has some plastic deformation, as shown in Figure 2.1b. When the material at the tip is yielding, the sharp tip of the crack "A" will become a dull tip, and the stress concentration factor decreases. Let assume that the stress concentration factor becomes 2.5. Now the maximum stress at the dull tip of the crack "A" is $2.5 \times 20 = 50$ ksi. It is less than the material yield strength. Therefore, the yielding at the tip of the crack "A" will stop. If external loading is not a cyclic loading, the effect of the microscopic crack "A" on the component is negligible and can be ignored because the plastic deformation at the tip of the crack "A" is in microscopic level.

2. Crack propagation. When the component is under a fully reversed cyclic loading, the reversed bending moment now tries to close the microscopic crack "A", as shown in Figure 2.1c. It is well known that if the dull tip of the microscopic crack "A" undergoes "strain hardening" due to the yielding [3], the material around the dull tip area will be brittle. After one time or a few times of such "open" and "close" actions, a new microscopic crack "B" will be generated beneath the crack "A". This result is a crack propagation, as shown in Figure 2.1c. When the component is under a fully reversed cyclic loading, the microscopic crack "B" will be opened again as shown in Figure 2.1d, which is the same situation as shown in Figure 2.1a. In this stage, the crack could continue to grow as a result of continuously applied cyclic loading.

3. Fracture due to static loading. Eventually, a crack will reach a critical size and the effective cross-sectional area of the component is so reduced that the actual stress on the effective

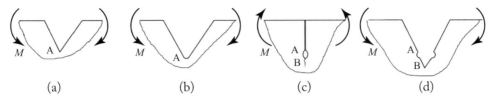

Figure 2.1: Schematic of a gradually crack propagation.

area under a normal service cyclic load will be larger than the ultimate material strength, causing the component to rupture due to static loading.

4. Fatigue damage is irreversible and will be gradually accumulated. The fatigue damage is due to crack propagation. When the cyclic loading stops, the propagated microscopic cracks still exist. Therefore, the fatigue damage is irreversible and will be gradually accumulated on the continuous cyclic loading.

2.3   FATIGUE TEST, S-N CURVE, AND MATERIAL ENDURANCE LIMIT

A cyclic load applied on a component can be any type of cyclic load and can be described by six models of cyclic loading spectrum [4], which has been described in Section 1.2. But, lots of material fatigue strength data is typically obtained from a stress-life method. In the stress-life method, a specimen is subjected to cyclic stress with a constant stress amplitude until it fractures and fails. There are many different types of fatigue specimen and fatigue test procedures. Fatigue test specimen will be designed and manufactured according to corresponding fatigue standards such as ASTM standards, and the test procedure will also follow the procedure defined by corresponding fatigue test standards. The cyclic stress for a fatigue test could be cyclic bending stress, cyclic axial stress, or cyclic shear (torsion) stress. The cyclic stress in a stress-life method is typically a fully reversed cyclic stress, that is, a constant stress amplitude with zero-mean stress. The main reasons for this are as follows. (1) Lots of fatigue test data are from rotating bending fatigue test, in which the cyclic stress is a fully reversed cyclic stress. (2) In fatigue theory for fatigue design, non-zero mean cyclic stress will typically be converted into fully reversed cyclic stress with an equivalent stress amplitude by including the effect of mean stress. (3) Even though fatigue tests are under cyclic stress with non-zero mean stress, it might be still presented as fatigue test data with an equivalently fully reversed cyclic stress for the purpose that the fatigue test data can be used for fatigue design. In the following, we will assume that cyclic stress in the stress-life method is a fully reversed cyclic stress.

In a stress-life method with a fatigue test specimen under a fully reversed cyclic stress, test results are the stress amplitude S'_f and the number of cycles at the failure N. Both S'_f and N is material fatigue strength data. This stress amplitude S'_f in a fatigue test is called as the material fatigue strength at the given number of cycles $n_L = N$. The physical meaning of this fatigue strength S_f' is that when the number of cycles of a fully reversed cyclic stress is $n_L = N$, the fatigue specimen will fail if the stress amplitude σ_a of a fully reversed cyclic stress is more than S_f'. In other words, the maximum stress amplitude of a fully reversed cyclic stress cannot exceed S_f' to avoid a fatigue failure when the service life is specified as $n = N$. The number of cycles at failure N in the fatigue test is called as the material fatigue life at this specified stress level $\sigma_a = S_f'$ of a fully reversed cyclic stress amplitude. The physical meaning of the material fatigue life N is that if the fatigue test specimen is under a fully reversed cyclic stress with a stress amplitude $\sigma_a = S_f'$, the fatigue specimen will fail when the service life n_L of such cyclic stress is more than N. Therefore, the fatigue test results (S_f', N) is a pair of fatigue strength data in a fatigue test.

After fatigue tests on the fatigue specimen of the same material are continuously conducted at different stress amplitudes (stress levels), a group data (S_f', N) will be collected and can be depicted as an S-N curve, as shown in Figure 2.2. The S-N curve is typically plotted in a Cartesian coordinate with a log-log scale.

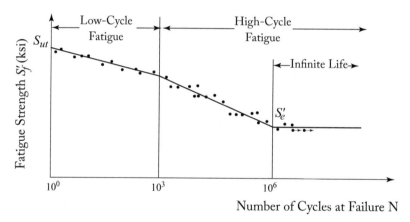

Figure 2.2: Schematic of an S-N curve.

In Figure 2.2, the small dots are a pair of fatigue test data. There are three different fatigue regimes, as shown in Figure 2.2. The fatigue failure form $N = 1$ to $N = 1000$ (cycles) is generally classified as low-cycle fatigue. In low-cycle fatigue, the stress level at the number of cycles at failure $N = 1$ is the ultimate material strength S_{ut}. For low-cycle fatigue, the test method is typically through strain-life method or linear-elastic fracture mechanics method, in which, the strain or the crack growth will be controlled or measured. This low-cycle fatigue is not the concerned topics of this book.

The fatigue failure with $N > 1000$ (cycles) is generally called as high-cycle fatigue. The high-cycle fatigue will be the focus of this book and is the typical case for most of the fatigue design in the industry.

Material endurance limit: For some materials like steels as shown in Figure 2.2, there is a value below which fatigue specimen will not fail with a very large number of cycles such as more than 10^6 (cycles). Material endurance limit S'_e is usually defined as the maximum fully reversed stress amplitude that a material can withstand infinitely without fracture. For some materials, the fatigue strength S'_f at the fatigue life $N = 10^6$ (cycles) is named as material endurance limit S'_e.

In high-cycle fatigue, a fatigue life between $N = 10^3$ (cycles) and $N = 10^6$ (cycles) is defined as a finite-life region, and a fatigue life $N \geq 10^6$ (cycles) is defined as an infinite life.

In a finite-life region, when there are fatigue tests on at least three different stress amplitude levels of fully reversed cyclic stress, the average fatigue life N at the same fatigue stress level vs. the fatigue strength S'_f in a log-log scale coordinate system can be typically simplified as a linear line, as shown in Figure 2.2. The material fatigue strength S'_f and the fatigue life N on this linear line has the following relationship:

$$N \left(S'_f \right)^m = Constant, \tag{2.1}$$

where S'_f and N are the material fatigue strength and the corresponding fatigue life on the simplified linear line. m is the slope of the traditional S-N curve and is a material mechanical property determined by fatigue test data. m can be determined through the linear least-squares regression by using the fatigue test results:

$$m = \frac{I \left[\sum_I (\ln \sigma_{ai} \cdot \ln N_i) \right] - \left(\sum_I \ln \sigma_{ai} \right) \left(\sum_I \ln N_i \right)}{I \left[\sum_I (\ln \sigma_{ai})^2 \right] - \left(\sum_I \ln \sigma_{ai} \right)^2}, \tag{2.2}$$

where I is the number of different stress amplitude levels σ_a for fatigue tests; σ_{ai} is the ith stress amplitude level in fatigue tests; $\ln N_i$ is the average fatigue life in a log-scale at the fatigue test level σ_{ai}. If there are a total J fatigue tests at the fatigue test stress level σ_{ai}, the $\ln N_i$ will be:

$$\ln N_i = \frac{\sum_J \ln N_{ij}}{J}, \tag{2.3}$$

where N_{ij} is the number of cycles at the failure of the jth fatigue test under the fatigue test stress level σ_{ai}.

When there are only a few fatigue tests for the S-N curve, Equation (2.1) is the design equation for the traditional fatigue design approach. The case with plenty of fatigue tests will be discussed in Sections 2.8 and 2.9 and will be the focus of this book.

Example 2.1
Fatigue tests of steel specimens under a fully reversed cyclic bending stress are listed in Table 2.1. Determine the material property m on a log-log scale.

Table 2.1: A group of fatigue test data

Stress Amplitude σ_a (Mpa)	Sample Size	Fatigue Life N (cycles) $\times 10^3$
392.40	4	34, 42, 43, 48
372.78	6	36, 47, 48, 53, 62, 65
353.16	6	60, 70, 77, 84, 89, 116
333.54	4	111, 114, 145, 197
313.92	5	171, 253, 254, 301, 309

Solution:

In this group of fatigue tests, there are five different stress amplitude levels. So, $I = 5$. In each fatigue stress levels, the fatigue tests are repeated for several times. We will use Equation (2.3) to calculate the average fatigue life in a log-scale at each stress amplitude level. For example, there are six repeated fatigue tests on the third stress level $\sigma_{a3} = 353.16$ (Mpa), the average fatigue life in a log scale $\log N_3$ will be:

$$\ln N_3 = \frac{\ln (60000) + \ln (70000) + \ln (77000) + \ln (84000) + \ln (89000) + \ln (116000)}{6}$$

$$= 11.30523.$$

The stress amplitudes and corresponding average fatigue life on a log-log scale for this example are listed in Table 2.2.

By using the data from Table 2.2 with Equation (2.2), the material property m is:

$$m = 8.303.$$

■

Table 2.2: Average log stress amplitudes and fatigue life

Stress level #	1	2	3	4	5
$\ln \sigma_{ai}$	5.972282	5.920989	5.866922	5.809763	5.749138
$\ln N_i$	10.63186	10.8372	11.30104	11.83417	12.43832

For fatigue design with an infinite fatigue life, the material endurance limit S'_e is one fatigue strength data for the material. When there is a lack of fatigue test data for material endurance limit S'_e, it can be estimated by using the ultimate strength of the same material [5].

For bending cyclic loading:

For steel,

$$S'_e = \begin{cases} 0.5 S_{ut} & S_{ut} < 1400 \text{ Mpa (200 ksi)} \\ 700 \text{ (Map) (100 ksi)} & S_{ut} \geq 1400 \text{ Mpa (200 ksi).} \end{cases} \tag{2.4}$$

For iron,

$$S'_e = \begin{cases} 0.4 S_{ut} & S_{ut} < 400 \text{ Mpa (60 ksi)} \\ 160 \text{ (Map) (24 ksi)} & S_{ut} \geq 400 \text{ Mpa (60 ksi).} \end{cases} \tag{2.5}$$

For aluminum,

$$S'_e = \begin{cases} 0.4 S_{ut} & S_{ut} < 330 \text{ Mpa (48 ksi)} \\ 130 \text{ (Map) (19 ksi)} & S_{ut} \geq 330 \text{ Mpa (48 ksi).} \end{cases} \tag{2.6}$$

For copper alloy,

$$S'_e = \begin{cases} 0.4 S_{ut} & S_{ut} < 280 \text{ Mpa (40 ksi)} \\ 100 \text{ (Map) (14 ksi)} & S_{ut} \geq 280 \text{ Mpa (40 ksi).} \end{cases} \tag{2.7}$$

For axial cyclic loading:
For steel,

$$S'_e = 0.45 S_{ut}. \tag{2.8}$$

For cyclic torsional loading:
For steel,

$$S'_e = 0.29 S_{ut}. \tag{2.9}$$

For iron,

$$S'_e = 0.32 S_{ut}. \tag{2.10}$$

For copper alloy,

$$S'_e = 0.22 S_{ut}. \tag{2.11}$$

2.4   THE MARIN MODIFICATION FACTORS

Material fatigue strength data are typically obtained from fatigue tests on standard rotating-beam bending specimen under fully reversed cyclic bending stress. The fatigue specimens are designed according to the fatigue test standards. For example, the rotating-beam bending stress specimen has a polish surface finish and a curved cylindrical shape with a smallest diameter 0.300″ in the middle of the specimen. The fatigue tests are typically under a fully reversed cyclic bending stress at the room temperature. For fatigue design, a component under consideration will have a different surface finish, different dimension, and different types of cyclic loading. Different surface finish will have quite different initial defects or cracks on the surfaces. A component with a bigger dimension means that it will have a much higher likelihood of more initial defects inside components. The maximum stress' area of a component due to bending, torsion, and axial loading are quite different. For a component under bending, the maximum stress will happen on the uppermost and lowermost layers. For a component under torsion, the maximum stress will appear on the outer surface. However, a component under axial loading, the maximum stress will appear on the whole cross-section. Therefore, component fatigue strength will be different from the material fatigue strength obtained from fatigue specimen tests. This difference of fatigue strength between fatigue test specimen and a component is typically considered by several Marin modification factors [2, 6, 7]. Those modifications on the material fatigue test data are based on the rotating-beam bending fatigue test under a fully reversed cyclic bending stress. The following equation can calculate component endurance limit S_e at the critical section:

$$S_e = k_a k_b k_c S_e', \tag{2.12}$$

where S_e' is the material endurance limit obtained from fatigue test on the fatigue test specimen. k_a is the surface finish modification factor. k_b is the size modification factor. k_c is the loading modification factor. A mechanical component might have several different component endurance limits at different critical section due to the different size modification factors.

Component fatigue strength S_f at a given fatigue life N can be obtained through the following equation:

$$S_f = k_a k_b k_c S_f', \tag{2.13}$$

where S_f' is material fatigue strength at the fatigue life N, which is the number of cycles at failure in the fatigue test. For fatigue design, the component fatigue strength S_f is not one value and will have a different value at different given fatigue life N. The rests in Equation (2.13) are the same as those in Equation (2.12).

The surface finish modification factor k_a can be treated as a normally distributed random variable. Its mean μ_{k_a} [7] will be calculated by the following equations:

$$\mu_{k_a} = \begin{cases} 16.45\,(S_{ut})^{-0.7427} & \text{For hot-rolled component} \\ 39.9\,(S_{ut})^{-0.995} & \text{For as-forged component} \\ 2.7\,(S_{ut})^{-0.2653} & \text{For machined surface component} \\ 1.34\,(S_{ut})^{-0.0848} & \text{For ground surface component,} \end{cases} \tag{2.14}$$

where S_{ut} is material ultimate tensile strength in the unit of ksi.

Its standard deviation σ_{k_a} will be calculated by using an estimated coefficient of variance γ_{k_a} [7] of the surface finish modification factor k_a by the following equations:

$$\gamma_{k_a} = \begin{cases} 0.098 & \text{For hot-rolled component} \\ 0.078 & \text{For as-forged component} \\ 0.06 & \text{For machined surface component} \\ 0.131 & \text{For ground surface component} \end{cases} \tag{2.15}$$

$$\sigma_{k_a} = \gamma_{k_a} \times \mu_{k_a}. \tag{2.16}$$

The size modification factor k_b will be treated as a deterministic and can be calculated by the following equation [7]:

$$k_b = \begin{cases} \left(\dfrac{d}{0.3}\right)^{-0.1133} & \text{For bending or torsion load with } 0.11'' \le d \le 2'' \\ 1 & \text{For axial load,} \end{cases} \tag{2.17}$$

where d is the diameter (or equivalent diameter) of the component in the unit of inch at the critical section.

The load modification factor k_c can be treated as a normally distributed random variable. Its mean μ_{k_c} can be calculated by the following equation [7]:

$$\mu_{k_c} = \begin{cases} 1 & \text{For bending load} \\ 0.774 & \text{For axial load} \\ 0.583 & \text{For torsional load.} \end{cases} \tag{2.18}$$

Its standard deviation σ_{k_c} will be calculated by using an estimated coefficient of variance γ_{k_c} [7] of the load modification factor k_c by the following equations:

$$\gamma_{k_c} = \begin{cases} 0 & \text{For bending load} \\ 0.163 & \text{For axial load} \\ 0.123 & \text{For torsional load} \end{cases} \tag{2.19}$$

$$\sigma_{k_c} = \gamma_{k_c} \times \mu_{k_c}. \tag{2.20}$$

In Equation (2.19), the coefficient of variance γ_{k_c} for bending loads is zero, and the mean value μ_{k_c} is 1. This result is because the fatigue strength test data comes from cyclic bending loading.

Example 2.2

A machined bar with a diameter 1.5″ is subjected to a cyclic torsion loading. Its ultimate material strength is 61.5 ksi. If the fatigue test data are obtained from rotating-beam specimen under fully reversed cyclic bending stress, determine the surface finish modification factor k_a, the size modification factor k_b, and the load modification factor k_c.

Solution:

The surface finish modification factor k_a will be treated as a normally distributed random variable. The mean μ_{k_a} of the surface finish modification factor k_a per Equation (2.14) is:

$$\mu_{k_a} = 2.7\,(S_{ut})^{-0.2653} = 2.7(61.5)^{-0.2653} = 0.9053. \tag{a}$$

The coefficient of variance of k_a per Equation (2.15) is:

$$\gamma_{k_a} = 0.06. \tag{b}$$

The standard deviation of k_a per Equation (2.16) is

$$\sigma_{k_a} = \gamma_{k_a} \times \mu_{k_a} = 0.06 \times 0.9053 = 0.0543. \tag{c}$$

The size modification factor k_b will be treated as a deterministic per Equation (2.17) is:

$$k_b = \left(\frac{d}{0.3}\right)^{-0.1133} = \left(\frac{1.5}{0.3}\right)^{-0.1133} = 0.8333. \tag{d}$$

The load modification factor k_c is treated as a normal distributed random variable. Its mean μ_{k_c} per Equation (2.18) is:

$$\mu_{k_c} = 0.583. \tag{e}$$

The coefficient of variance of $\gamma_{k_{ac}}$ per Equation (2.19) is:

$$\gamma_{k_c} = 0.123. \tag{f}$$

The standard deviation of k_c per Equation (2.20) is

$$\sigma_{k_c} = \gamma_{k_c} \times \mu_{k_c} = 0.123 \times 0.583 = 0.0717. \tag{g}$$

∎

2.5   THE EFFECT OF MEAN STRESS

Fatigue strength data is typically from fatigue tests under a fully reversed cyclic stress. Even when a fatigue test is under a non-zero-mean cyclic stress, it is typically presented as a fatigue strength data with an equivalent stress amplitude of a fully reversed cyclic stress. This approach is simply because general cyclic loading for mechanical component fatigue design might be any non-zero-mean stress cyclic stress. There are many fatigue theories such as Soderberg approach, Modified Goodman approach, Gerber approach, and ASME-Elliptic approach for considering the effect of mean stress [2, 5]. This book will use the Modified Goodman approach to consider the effect of mean stress in cyclic stress through the following equation:

$$\sigma_{a-eq} = \begin{cases} \sigma_a \left(\dfrac{S_{ut}}{S_{ut} - \sigma_m} \right) & \text{when } \sigma_m \geq 0 \\ \sigma_a & \text{when } \sigma_m < 0, \end{cases} \tag{2.21}$$

where σ_a and σ_m are the stress amplitude and the mean stress of cyclic stress. In Equation (2.21), S_{ut} is the ultimate material strength as a deterministic value, which will be equal to the average value of the ultimate material strength. σ_{a-eq} is the equivalent stress amplitude of a fully reversed cyclic stress. For cyclic stress with negative mean stress, the equivalent stress amplitude will be equal to the stress amplitude of the cyclic stress with negative mean stress because the compressed mean stress will help to stop the crack propagation. Therefore, the Modified Goodman approach is more conservative.

Example 2.3
A component is subjected to cyclic stress with a mean stress $\sigma_m = 5$ (ksi) and a stress amplitude $\sigma_a = 17$ (ksi). The ultimate material strength is 61.5 (ksi). Determine its equivalent stress amplitude of a fully reversed cyclic stress.

Solution:
Per Equation (2.21), the equivalent stress amplitude in this example with a positive mean stress $\sigma_m = 5$ (ksi) will be:

$$\sigma_{a-eq} = \sigma_a \left(\frac{S_{ut}}{S_{ut} - \sigma_m} \right) = 17 \left(\frac{61.5}{61.5 - 5} \right) = 18.5 \ (\text{ksi}).$$

∎

Example 2.4
A fatigue test specimen is under cyclic stress with a mean stress $\sigma_m = 15$ (ksi) and a stress amplitude $\sigma_a = 15$ (ksi). The fatigue life, that is, the number of cycles at failure under such cyclic loading for this fatigue specimen is 6.5×10^5 (cycles). The ultimate strength of the material of the specimen is 61.5 (ksi). Express this fatigue test data as a fatigue test data under a fully reversed cyclic stress.

Solution:
Per Equation (2.21), the equivalent fatigue strength in this example with a positive mean stress $\sigma_m = 21$ (ksi) will be:

$$S'_f = \sigma_{a-eq} = \sigma_a \left(\frac{S_{ut}}{S_{ut} - \sigma_m} \right) = 21 \left(\frac{61.5}{61.5 - 21} \right) = 31.89 \ (\text{ksi}).$$

So, for this fatigue test, the fatigue test results could be equivalently expressed by:

- the fatigue strength $S_f' = 31.89$ ksi at the fatigue life $N = 6.5 \times 10^5$ (cycles) under a fully reversed cyclic stress, that is, ($S_f' = 31.89$ ksi, $N = 6.5 \times 10^5$ cycles).

■

2.6   THE FATIGUE STRESS CONCENTRATION FACTOR

The fatigue stress-concentration factor K_f will be used to multiply the nominal stress amplitude and can be treated as a normally distributed random variable. Its mean μ_{K_f} can be calculated by the following equation [7]:

$$\mu_{K_f} = \frac{K_t}{1 + \dfrac{2}{\sqrt{r}} \left(\dfrac{K_t - 1}{K_t} \right) \sqrt{a}}, \qquad (2.22)$$

where K_t is static stress concentration factor which can be obtained through any design handbook and some websites. r is the notch radius in the unit of inch. $\sqrt{a}$ is defined as the Neuber constant and can be calculated through the following equation:

$$\sqrt{a} = \begin{cases} \dfrac{5}{S_{ut}} & \text{For a transverse hole} \\ \dfrac{4}{S_{ut}} & \text{For a shoulder} \\ \dfrac{5}{S_{ut}} & \text{For a groove,} \end{cases} \qquad (2.23)$$

where S_{ut} is material tensile ultimate strength in the unit of ksi.

The coefficient of variance of the fatigue stress concentration factor K_f can be estimated by the following equation:

$$\gamma_{K_f} = \begin{cases} 0.11 & \text{For a transverse hole} \\ 0.08 & \text{For a shoulder} \\ 0.13 & \text{For a groove.} \end{cases} \qquad (2.24)$$

The standard deviation of the fatigue stress concentration factor K_f will be:

$$\sigma_{K_f} = \mu_{K_f} \times \gamma_{K_f}. \qquad (2.25)$$

Example 2.5
A machined steel shaft with a shoulder, as shown in Figure 2.3 is subjected to cyclic bending stress. The ultimate strength of the shaft material is 61.5 (ksi). Determine the fatigue stress concentration factor K_f in the shoulder of the shaft.

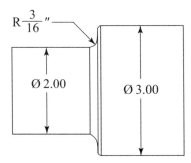

Figure 2.3: Schematic of a segment of a shaft with a shoulder.

Solution:
The Neuber constant on the shaft shoulder per Equation (2.23) is:

$$\sqrt{a} = \frac{4}{S_{ut}} = \frac{4}{61.5} = 0.06504. \tag{a}$$

From design handbooks or some websites about static stress concentration factors, the theoretical geometric stress concentration of this shoulder K_t due to bending is

$$K_t = 1.78. \tag{b}$$

The mean μ_{K_f} of the fatigue stress concentration factor K_f in the shoulder of the shaft per Equation (2.22) is

$$\mu_{K_f} = \frac{K_t}{1 + \dfrac{2}{\sqrt{r}}\left(\dfrac{K_t - 1}{K_t}\right)\sqrt{a}} = \frac{1.78}{1 + \dfrac{2}{\sqrt{0.1875}}\left(\dfrac{1.78 - 1}{1.78}\right)0.06504} = 1.573. \tag{c}$$

The standard deviation of the fatigue stress concentration factor K_f per Equations (2.24) and (2.25) is:

$$\sigma_{K_f} = \mu_{K_f} \times \gamma_{K_f} = 1.573 \times 0.08 = 0.1258. \tag{d}$$

∎

2.7 RELIABILITY OF A COMPONENT WITH AN INFINITE LIFE

For a component with infinite life, the component endurance limit S_e will be used for the component fatigue design. The limit state function of a component with an infinite life will be:

$$g\left(S_e, K_f, \sigma_{eq-a}\right) = S_e - K_f \times \sigma_{eq-a} = \begin{cases} > 0 & \text{Safe} \\ 0 & \text{Limit state} \\ < 0 & \text{Failure,} \end{cases} \qquad (2.26)$$

where S_e is the component endurance limit at the critical section, which is defined by Equation (2.12). K_f is the fatigue stress concentration factor at the critical section and is defined by Equations (2.22)–(2.25). σ_{eq-a} is the component's equivalent fully reversed cyclic stress amplitude at the critical section and is determined by Equation (2.21).

The limit state function Equation (2.26) can be used to calculate the reliability of a component with infinite life. The H-L, R-F, or Monte Carlo method, which were discussed in Chapter 3 of Le [8], can be used to calculate its reliability. The concise description of procedures of three methods is presented in Appendix A of this book. We will use three examples to demonstrate how to calculate the reliability of a component with infinite life. The corresponding MATLAB programs will be displayed in Appendix B for a reference.

Example 2.6

A machined constant circular bar with a diameter $d = 1.250 \pm 0.005''$ is subjected a cyclic axial loading. The mean axial loading F_m of the cyclic axial loading is a constant and equal to 12 (klb). The loading amplitude F_a of the cyclic loading follows a normal distribution with a mean $\mu_{F_a} = 8.5$ (klb) and a standard deviation $\sigma_{F_a} = 1.2$ (klb). The ultimate material strength is 61.5 (ksi). Its endurance limit S_e' follows a normal distribution with a mean $\mu_{S_e'} = 24.7$ (ksi) and a standard deviation $\sigma_{S_e'} = 2.14$ (ksi), which are based on fatigue tests under fully reversed bending stress. This bar is designed to have an infinite life. (1) Establish the limit state function of this problem. (2) Calculate the reliability of the bar under the cyclic axial loading.

Solution:

(1) Establish the limit state function of this problem.

We can use Equation (2.26) to establish the limit state function for this problem. Per Equation (2.12), the component endurance limit will be:

$$S_e = k_a k_b k_c S_e'. \qquad (a)$$

Per Equations (2.14), (2.15), and (2.16), the surface finish modification factor k_a of the machined bar will be a normal distribution with the following distributed parameters:

$$\mu_{k_a} = 2.7\,(S_{ut})^{-0.2653} = 2.7 \times 61.5^{-0.2653} = 0.905 \tag{b}$$

$$\sigma_{k_a} = \gamma_{k_a} \times \mu_{k_a} = 0.06 \times 0.905 = 0.0543. \tag{c}$$

Per Equation (2.17), the size modification factor k_b will be treated as a constant and is equal to 1 due to the cyclic axial loading:

$$k_b = 1. \tag{d}$$

Per Equations (2.18), (2.19), and (2.20), the loading modification factor k_c will be treated as a normal distribution with the following mean and standard deviation:

$$\mu_{k_c} = 0.774 \tag{e}$$

$$\sigma_{k_c} = \gamma_{k_c} \times \mu_{k_c} = 0.774 \times 0.163 = 0.1262. \tag{f}$$

For this problem, the fatigue stress concentration factor K_f will be equal to 1 due to a constant circular cross-section.

The mean stress σ_m and the stress amplitude σ_a of the cyclic axial stress in this problem can be calculated by the following equations:

$$\sigma_m = \frac{F_m}{\pi d^2/4} = \frac{4F_m}{\pi d^2} = \frac{4 \times 12}{\pi d^2} = \frac{15.279}{d^2} \text{ (ksi)} \tag{g}$$

$$\sigma_a = \frac{F_a}{\pi d^2/4} = \frac{4F_a}{\pi d^2} = \frac{1.273F_a}{d^2} \text{ (ksi).} \tag{h}$$

The diameter d of the bar can be treated as a normal distribution. Its mean μ_d and standard deviation σ_d can be determined per Equation (1.1):

$$\mu_d = 1.250, \qquad \sigma_d = \frac{0.005 - (-0.005)}{8} = 0.00125. \tag{i}$$

Since the cyclic axial stress is not a fully reversed cyclic stress, we need to use Equation (2.21) to consider the effect of mean stress and converted it into a fully reversed cyclic stress with an equivalent stress amplitude:

$$\sigma_{a-eq} = \sigma_a \left(\frac{S_{ut}}{S_{ut} - \sigma_m}\right) = \frac{1.273F_a}{d^2}\left(\frac{61.5}{61.5 - \dfrac{15.279}{d^2}}\right) = \frac{78.290F_a}{61.5d^2 - 15.279}. \tag{j}$$

Now, by using all information from Equations (a)–(j), we can establish the limit state function for this problem per Equation (2.26):

$$g\left(k_a, k_c, S'_e, d, F_a\right) = k_a k_c S'_e - \frac{78.290F_a}{61.5d^2 - 15.279}. \tag{k}$$

Table 2.3: The distribution parameters of random variables in Equation (k)

k_a		k_c		S'_e (ksi)		d (in)		F_a (klb)	
μ_{k_a}	σ_{k_a}	μ_{k_c}	σ_{k_c}	$\mu_{S'_e}$	$\sigma_{S'_e}$	μ_d	σ_d	μ_{F_a}	σ_{F_a}
0.905	0.0543	0.774	0.1262	24.7	2.14	1.250	0.0125	8.5	1.2

In this example, all random variables in the limit state function (k) are normal distributions. Their distribution parameters are listed in Table 2.3.

(2) Use the H-L method to calculate the reliability of the bar.

The limit state function (k) contains five normally distributed random variable and is a nonlinear function. We will follow the H-L method presented in Appendix A.1 and the program flowchart in Figure A.1 to create a MATLAB program. This MATLAB program is displaced in Appendix B as "B.1: The H-L method for Example 2.6."

The iterative results are listed in Table 2.4. From the iterative results, the reliability index β and corresponding reliability R of the bar in this example are:

$$\beta = 2.792757 \qquad R = \Phi(2.792757) = 0.9974.$$

∎

Table 2.4: The iterative results of Example 2.6 by the H-L method

Iterative #	k_a^*	k_c^*	S'^*_e	d^*	F_a^*	β^*	$|\Delta\beta^*|$
1	0.905	0.774	24.7	1.25	17.85956	2.534002	
2	0.865081	0.52188	22.42825	1.246358	10.37987	2.800277	0.266275
3	0.872351	0.481666	22.74402	1.247141	9.811213	2.793018	0.00726
4	0.874674	0.477329	22.8934	1.247324	9.816224	2.792766	0.000252
5	0.874928	0.476348	22.91546	1.24734	9.808635	2.792757	8.78E-06

Example 2.7

The critical section for a machined rotating shaft is on the shoulder section, as shown in Figure 2.4. The bending moment M (klb.in) on the shoulder section can be described by a lognormal distribution with a log-mean $\mu_{\ln M} = 0.315$ and a log-standard deviation $\sigma_{\ln M} = 0.142$. The shaft material's ultimate strength is 61.5 (ksi). Its endurance limit S'_e follows a normal distribution with a mean $\mu_{S'_e} = 24.7$ (ksi) and a standard deviation $\sigma_{S'_e} = 2.14$ (ksi), which are based on fatigue tests under fully reversed bending stress. This shaft is designed to have an infinite life. (1) Establish the limit state function of this shaft. (2) Calculate the reliability of the shaft.

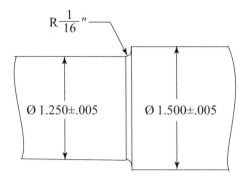

Figure 2.4: Schematic of a shoulder section of a shaft.

Solution:

(1) Establish the limit state function of the rotating shaft.

We can use Equation (2.26) to establish the limit state function for this problem. Per Equation (2.12), the component endurance limit will be:

$$S_e = k_a k_b k_c S_e'. \tag{a}$$

The mean and standard deviation of the surface finish modification factor k_a of the machined shaft can be calculated per Equations (2.14), (2.15), and (2.16):

$$\mu_{k_a} = 0.905; \qquad \sigma_{k_a} = 0.0543. \tag{b}$$

The size modification factor k_b can be calculated per Equation (2.17):

$$k_b = 0.8507. \tag{c}$$

The loading modification factor k_c will be 1 because the cyclic stress is cyclic bending stress:

$$k_c = 1. \tag{d}$$

The static stress concentration K_t in this case, can be obtained from any design handbook or a stress concentration website:

$$K_t = 1.96. \tag{e}$$

The mean and standard deviation of the fatigue stress concentration factor K_f can be calculated per Equations (2.22)–(2.25):

$$\mu_{K_f} = 1.562; \qquad \sigma_{K_f} = 0.1250. \tag{f}$$

For a rotating shaft, the bending moment will induce a fully reversed cyclic bending stress. Its stress amplitude of the fully reversed cyclic bending stress will be:

$$\sigma_a = \frac{M \times \dfrac{d}{2}}{I} = \frac{M \times \dfrac{d}{2}}{\dfrac{\pi}{64} d^4} = \frac{32M}{\pi d^3}. \tag{g}$$

The diameter d of the shaft can be treated as a normal distribution. Its mean μ_d and standard deviation σ_d can be determined per Equation (1.1):

$$\mu_d = 1.250, \qquad \sigma_d = \frac{0.005 - (-0.005)}{8} = 0.00125. \tag{h}$$

Now, by using all information from (a)–(h), we can establish the limit state function for this problem per Equation (2.26):

$$g\left(M, k_a, S'_e, K_f, d\right) = k_a k_b S'_e - K_f \frac{32M}{\pi d^3} = \begin{cases} > 0 & \text{Safe} \\ 0 & \text{Limit state} \\ < 0 & \text{Failure.} \end{cases} \tag{i}$$

In this example, M is a lognormal distribution, and the rest random variables in the limit state function (i) are normal distributions. Their distribution parameters are listed in Table 2.5.

Table 2.5: The distribution parameters of random variables in Equation (i)

M (klb.in), Lognormal Distribution		k_a, Normal Distribution		S'_e (ksi), Normal Distribution		K_f, Normal Distribution		d (in), Normal Distribution	
$\mu_{\ln M}$	$\sigma_{\ln M}$	μ_{k_a}	σ_{k_a}	$\mu_{S'_e}$	$\sigma_{S'_e}$	μ_{K_f}	σ_{K_f}	μ_d	σ_d
0.315	0.142	0.905	0.0543	24.7	2.14	1.562	0.1250	1.250	0.00125

(2) Use the R-F method to calculate the reliability of the shaft.

The limit state function (i) contains four normally distributed random variable and one lognormal distribution variable. We will use the R-F method to calculate the reliability of this example. The procedure of the R-F method and the program flowchart are presented in Appendix A.2. The MATLAB program of the R-F method for this example is displayed in Appendix B as "B.2: The R-F method for Example 2.7."

The iterative results are listed in Table 2.6. From the iterative results, the reliability index β and corresponding reliability R of the bar in this example are:

$$\beta = 2.72094 \qquad R = \Phi(2.72094) = 0.9968.$$

∎

Table 2.6: The iterative results of Example 2.6 by the H-L method

Iterative #	M^*	k_a^*	S'^*_e	k_f^*	d^*	β^*	$|\Delta\beta^*|$
1	1.384144	0.905	24.7	1.562	1.050139	2.993	
2	1.800485	0.85477	21.84149	1.716223	1.256083	2.72047	0.272531
3	1.808252	0.85755	21.81578	1.687236	1.249902	2.720937	0.000468
4	1.807202	0.857805	21.81853	1.689117	1.249948	2.72094	2.67E-06

Example 2.8

The critical section of a constant rectangular cross-section beam with a height $h = 2.00 \pm 0.010''$ and a width $b = 2.0 \pm 0.010''$ is under a cyclic bending loading. This cyclic bending loading has a constant mean bending moment $M_m = 20.5$ (klb.in). Its bending moment amplitude can be treated as a normal distribution with a mean $\mu_{M_a} = 11.5$ (klb.in) and a standard deviation $\sigma_{M_a} = 1.5$ (klb.in). The beam can be treated as a hot-rolled component. The beam material's ultimate strength is 61.5 (ksi). Its endurance limit S'_e follows a normal distribution with a mean $\mu_{S'_e} = 24.7$ (ksi) and a standard deviation $\sigma_{S'_e} = 2.14$ (ksi), which are based on fatigue tests under fully reversed bending loading. This beam is designed to have an infinite life. (1) Establish the limit state function of this beam at the critical cross-section. (2) Calculate the reliability of the beam by using the Monte Carlo method, the relative error and the reliability range with a 95% confidence level.

Solution:

(1) Establish the limit state function of this beam at the critical cross-section.

We can use Equation (2.26) to establish the limit state function for this problem. Per Equation (2.12), the component endurance limit will be:

$$S_e = k_a k_b k_c S'_e. \tag{a}$$

The mean and standard deviations of the surface finish modification factor k_a of the hot-rolled component can be calculated per Equations (2.14), (2.15), and (2.16):

$$\mu_{k_a} = 0.772; \qquad \sigma_{k_a} = 0.0757. \tag{b}$$

Per Equation (2.17), the size modification factor k_b will be treated as a constant. Since the cross-section is not a circular cross-section and the beam is not a rotating component, the equivalent diameter should be used for the calculation. The equivalent diameter [2] for non-rotating rectangular cross-section for this case is $d_e = 0.808\sqrt{bh} = 0.808\sqrt{2 \times 2} = 1.616''$. So, the size modification factor k_b can be calculated per Equation (2.17):

$$k_b = \left(\frac{d}{0.3}\right)^{-0.1133} = \left(\frac{1.616}{0.3}\right)^{-0.1133} = 0.826. \qquad (c)$$

Per Equations (2.18), (2.19), and (2.20), the loading modification factor k_c will be treated as a constant 1 due to the cyclic bending stress:

$$k_c = 1. \qquad (d)$$

For this problem, the fatigue stress concentration factor K_f will be equal to 1 due to a constant cross-section.

The mean stress σ_m and the stress amplitude σ_a on the critical cross-section due to the cyclic bending moment can be calculated by the following equations:

$$\sigma_m = \frac{M_m \frac{h}{2}}{bh^3/12} = \frac{6F_m}{bh^3} = \frac{6 \times 20.5}{bh^2} = \frac{123}{bh^2} \text{ (ksi)} \qquad (e)$$

$$\sigma_a = \frac{M_a \frac{h}{2}}{bh^3/12} = \frac{6M_a}{bh^2} \text{ (ksi)}. \qquad (f)$$

The height h and the width b of the beam can be treated as normally distributed random variables. Their means and standard deviations can be determined per Equation (1.1):

$$\mu_h = 2.0, \qquad \sigma_h = \frac{0.010 - (-0.010)}{8} = 0.0025 \qquad (g)$$

$$\mu_b = 1.0, \qquad \sigma_b = \frac{0.010 - (-0.010)}{8} = 0.0025. \qquad (h)$$

Since the cyclic bending stress is not a fully reversed cyclic stress, we need to use Equation (2.21) to consider the effect of mean stress and to convert it into a fully reversed cyclic stress with an equivalent stress amplitude:

$$\sigma_{a-eq} = \sigma_a \left(\frac{S_{ut}}{S_{ut} - \sigma_m}\right) = \frac{6M_a}{bh^2} \left(\frac{61.5}{61.5 - \frac{123}{bh^2}}\right) = \frac{369M_a}{61.5bh^2 - 123}. \qquad (i)$$

Now, by using all information from (a)–(i), we can establish the limit state function for this problem per Equation (2.26):

$$g\left(k_a, S_e', h, b, M_a\right) = 0.826k_a S_e' - \frac{369M_a}{61.5bh^2 - 123}. \qquad (j)$$

In this example, all random variables in the limit state function (j) are normal distributions. Their distribution parameters are listed in Table 2.7.

Table 2.7: The distribution parameters of random variables in Equation (j)

k_a		S_e' (ksi)		h (in)		b (in)		M_a (klb)	
μ_{k_a}	σ_{k_a}	$\mu_{S_e'}$	$\sigma_{S_e'}$	μ_h	σ_h	μ_b	σ_d	μ_{M_a}	σ_{M_a}
0.772	0.0757	24.7	2.14	2.0	0.00025	2.0	0.00025	11.5	1.5

(2) Use the Monte Carlo method to calculate the reliability of the beam.

The limit state function (j) contains five normally distributed random variable and is a nonlinear function. We can follow the procedure and the flowchart of the Monte Carlo method presented in Appendix A.3 to compile a MATLAB program. It is displayed in Appendix B as "B.3: The Monte Carlo method for Example 2.8." The program results are:

The reliability of the component: $R = 15280173/15998400 = 0.9551$.

The failure probability of the component: $F = 1 - R = 0.00459$.

The relative error of the failure probability: $\varepsilon = 0.00231$.

The error range of the failure probability: $F = F \pm \varepsilon F = 0.00459 \pm 0.0001$.

The reliability range with a 95% confidence level: $R = 0.9551 \pm 0.0001$.   ■

2.8 RELIABILITY OF A COMPONENT BY THE P-S-N CURVES APPROACH

2.8.1 THE MATERIAL P-S-N CURVES

The fatigue failure of a component is mainly due to the "defects" such as manufactured scratches on the surfaces of a component, "dislocations," "impurity," "micro-cracks," or "micro-cavities" inside a component. The "defects" or the "uncertainty" are randomly scattered on the surface of or inside the component. Therefore, fatigue strengths of the component under cyclic loading due to "uncertainty" are random variables. Fatigue tests usually cost lots of time and human resources. If there are only a few fatigue tests, the traditional S-N is used for fatigue design, which has been briefly in Section 2.3. When there are at least three different stress amplitude levels with enough number (at least 30) of fatigue tests at each cyclic stress level, the P-S-N curves can be used to describe material fatigue strength. The P-S-N curves approach is a well-known approach for describing the uncertainty in fatigue strength. The P-S-N curves approach, as shown in Figure 2.5, is a common probabilistic fatigue design theory for components under cyclic loading spectrum.

The P-S-N curves contain two sets of distribution functions: (1) the distribution functions of the failure cycle number at different constant cyclic stress amplitudes. These functions are also called as the P-N distributions (Probabilistic-Number of the cycle) at the given fatigue strength S_f'; and (2) the distribution functions of cyclic stress amplitudes at different constant cycle numbers. These functions are also named as the P-S distribution (Probabilistic-fatigue Strength) at

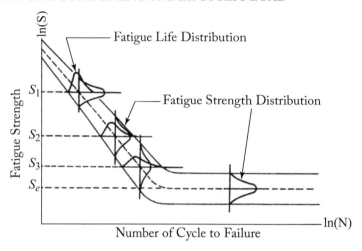

Figure 2.5: A schematic sketch of the P-S-N curve.

the given fatigue life N (the number of cycles to failure). These two sets of distribution functions can be used for presenting fatigue test data and estimating the reliability of a component under cyclic loading.

The distribution functions of the failure cycle number at different constant cyclic stress amplitudes can be directly obtained through the fatigue tests under the same cyclic stress level (fatigue strength) and can usually be described by a lognormal distribution [7, 9].

The distribution functions of fatigue strength at different constant fatigue life can usually be described by the three parameters Weibull distribution or normal distribution or lognormal distribution, and its distribution parameters are derived by the statistical S-N envelope as shown in Figure 2.5 [7, 9].

Following, we will show two examples of the distributions of the fatigue life at the given fatigue strength, that is, the P-N distributions at the given fatigue strength S'_f.

In 1969, Dr. Dimitri B. Kececioglu and his colleagues presented P-S-N curves of several fatigue test data [10]. One set of fatigue date is the P-N curves of a carbon cold-drawn steel wire specimen under fully reversed cyclic bending loading, as shown in Table 2.8 [10, 11]. The numbers of fatigue tests in the first two stress levels are more than 30. The number of fatigue tests in the third level is almost 30. The number of fatigue tests at rests of stress levels are less than 30. Table 2.8 shows five lognormal distributed material fatigue life at the corresponding material fatigue strength (fully reversed stress amplitude) of the P-N curves. For example, at the given fatigue strength $S'_f = 66$ (ksi), the corresponding material fatigue life N is a lognormal distributed random variable with a log mean $\mu_{\log N} = 5.587$ and a log standard deviation $\sigma_{\log N} = 0.108$. This distribution can be used for reliability-based fatigue design.

Table 2.8: P-S-N curves of a steel wire specimen under fully reversed cyclic bending stress [10, 11]

Fatigue Strength (fully reversed stress amplitude ksi)	Number of Specimens	Lognormal Distributed the Number of Cycles to Failure	
		Mean μ_{logN}	Standard Deviation σ_{logN}
66	50	5.587	0.108
76	37	5.140	0.094
86	26	4.715	0.068
96	17	4.394	0.052
106	10	4.102	0.073

During 2016–2018, we used the hydraulic Instron 8801 fatigue test machine to conduct a total of 195 fatigue tests under cyclic axial loading [12]. The hydraulic Instron 8801 fatigue test machine, as shown in Figure 2.6, is a standard device manufactured by Instron for running fatigue tests under cyclic axial loading. The fatigue test specimen is clamped by the upper grip and the lower grip. The upper grip is connected to the load cell. The lower grip is directly connected with a hydraulic actuator. It forces the lower grip to have a cyclic up-and-down motion.

Figure 2.6: A photo of Instron 8801 fatigue test machine.

The fatigue test specimen was made from Aluminum 6061-T6 10 Gauge sheet. The chemical compositions of this 10 gauge 6061-T6 sheet provided by the material supplier are displayed in Table 2.9. The mechanical properties of this sheet provided by the material supplier are listed in Table 2.10.

Table 2.9: Chemical composition of the 6016-T6 sheet

Element	Si	Fe	Cu	Mn	Mg	Cr	Zn	Ti	Others	Al
%	0.629	0.306	0.276	0.020	1.030	0.185	0.004	0.021	0.15	Remain

Table 2.10: Mechanical properties of the sheet

Size	0.100″X48″X144″	Tensile strength	51.2 ksi	Yield strength	41.8 ksi	Elongation %	16

The fatigue specimen was a rectangular sheet-type flat fatigue test specimen, shown in Figure 2.8, and was designed according to ASTM STM E466–15, Standard Practice for Conducting Force Controlled Constant Amplitude Axial Fatigue Tests of Metallic Materials [13]. The manufacturing route for this specimen is: (1) use a sheet shearing machine to shear the sheet into a rectangular plate 11.75″ × 2.375″ along the longitudinal direction; (2) use a milling machine to mill the sheared rectangular plate into a rectangular plate 11.375″ × 2″; and (3) clamp the milled rectangular plate in the designed fixture and then use the compiled CNC program to mill the plate per the drawing shown in Figure 2.7.

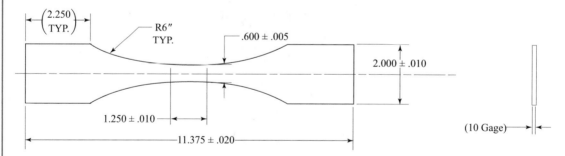

Figure 2.7: The dimensions of the sheet-type flat fatigue specimen.

Since the thickness of the sheet-type fatigue specimen is only 0.100″, the fatigue test specimen under cyclic axial loading is very easy to be buckling. To avoid the buckling during the cyclic axial loading, the cyclic axial loading for the fatigue test was with a loading ratio $S_r = 0$, that is, the axial loading varying between 0 to maximum tensile loading. The fatigue tests under five different stress levels with a total 195 of fatigue tests have been conducted. The fatigue test procedure is: (1) visually check the fatigue test specimen before testing. Any abnormal fatigue specimen such as a bent specimen or visual big crack or scratch on the outer surfaces of the

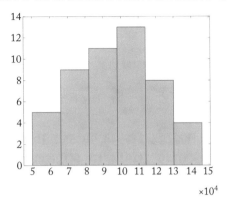

Figure 2.8: Histogram for the cyclic stress level: $\sigma_a = 20.833$ (ksi) and $\sigma_m = 20.833$ (ksi).

specimen will be discarded; (2) measure the actual width and thickness of the specimen and record them in a test log; (3) install specimen and make sure that the specimen is installed vertically and centrally in the upper and lower grips; (4) use the compiled WaveMatrix program to run the cyclic axial loading fatigue test, which is the control program for the Instron 8801 fatigue test machine, until the specimen is fractured; and (5) record the number of cycles at failure and some special notes in the test log.

For all following fatigue tests, the dimensions of the middle section of the specimen are the width $b = 0.600 \pm 0.005''$ and the thickness $t = 0.100 \pm 0.005''$. The loading frequency is 20 (Hz); that is, the cyclic loading cycles will be repeated 20 times in a second. The stress ratio $S_r = \frac{F_{\min}}{F_{\max}} = 0$. The test conditions and results of these five different stress levels are listed in Table 2.11. Stresses in Table 2.11 are calculated by using the nominal dimensions, that is, the width $b = 0.600''$ and the thickness $t = 0.100''$.

In Table 2.11, F_a and F_m are the loading amplitude and the loading mean of the cyclic axial loading. σ_a and σ_m are the stress amplitude and the stress mean of the cyclic axial stress under the cyclic axial loading with F_a and F_m. N is the number of cycles to failure of each fatigue specimen test under the corresponding specified cyclic loading.

The minimum $N_{\min}$, the maximum $N_{\max}$, the mean μ_N, and the standard deviation σ_N of the number of cycles to failure under five different stress levels are listed in Table 2.12. From Table 2.12, the number of cycles to failure at the same stress level has a very big variation. For example, at the cycling loading level $F_a = 1325$ (lb) and $F_m = 1325$ (lb), the maximum cycle of numbers to failure is almost three times of the minimum cycle of the number to failure.

The number of fatigue tests in each stress level listed in Table 2.11 is more than 30. Histograms of the number of cycles to failure N in each stress level are displayed in Figures 2.8–2.12. Figure 2.8 is the histogram of N with a sample size 50 at the first stress level $F_a = F_m = 1250$ (lb). Figure 2.9 is the histogram of N with a sample size 55 at the second

Table 2.11: Fatigue test data under cyclic axial loading with a stress ratio $S_r = 0$ [11]

F_a (lb)	F_m (lb)	σ_a (ksi)	σ_m (ksi)	Frequency	Samples
1250	1250	20.833	20.833	20 (Hz)	50
N	92171, 70272, 81285, 122628, 117940, 108584, 105626, 108478, 91504, 124341, 82764, 72315, 61440, 70179, 111914, 105461, 63454, 77811, 112188, 97636, 88818, 125856, 125108, 128583, 83833, 131692, 129422, 91333, 65294, 88960, 112093, 106840, 107792, 76654, 80709, 144728, 90767, 109190, 119484, 107040, 76279, 134353, 61451, 91384, 112944, 78865, 90895, 59056, 133075, 107914				
F_a (lb)	F_m (lb)	σ_a (ksi)	σ_m (ksi)	Frequency	Samples
1325	1325	22.083	22.083	20 (Hz)	55
N	75819, 65144, 74538, 57606, 56898, 83130, 70734, 81732, 59123,34609, 51723, 67689, 90634, 92805, 80232, 80146, 50395, 58294, 58306, 77666, 45867, 53352, 45295, 46894, 74711, 66332, 67625, 45168, 65383, 52614, 59240, 81547, 51968, 53924, 63214, 82222, 67763, 71004, 79390, 56301, 72795, 86831, 45676, 59665, 64815, 47176, 63627, 61775, 31474, 62380, 76340, 62633, 55449, 64220, 62850,				
F_a (lb)	F_m (lb)	σ_a (ksi)	σ_m (ksi)	Frequency	Samples
1350	1350	22.5	22.5	20 (Hz)	30
N	55740, 30958, 61312, 50246, 69952, 69652, 44155, 75026, 69953, 76280, 42858, 81745, 54140, 62791, 83492, 58030, 67268, 64992, 54181, 60379, 41778, 71907, 68426, 52853, 46879, 56964, 51692, 62875, 37836, 40000				
F_a (lb)	F_m (lb)	σ_a (ksi)	σ_m (ksi)	Frequency	Samples
1375	1375	22.917	22.917	20 (Hz)	30
N	60639, 56861, 45306, 50836, 46038, 56269, 54201, 50630, 67016, 46932, 31499, 67836, 33708, 53025, 52895, 42112, 49586, 30872, 46168, 55442, 51839, 40660, 31760, 45480, 41600, 50742, 64874, 34599, 67186, 53949				
F_a (lb)	F_m (lb)	σ_a (ksi)	σ_m (ksi)	Frequency	Samples
1400	1400	23.333	23.333	20 (Hz)	30
N	57008, 35051, 32983, 30485, 39212, 68979, 33110, 60411, 54700,50617, 33079, 57704, 44198, 37292, 47345, 45435, 40340, 64051, 51288, 50667, 46380,55792, 45103, 44285, 37200, 31247, 30815, 48816, 50164, 66793				

Table 2.12: Some statistical values about the number of cycles to failure N

F_a (lb)	F_m (lb)	N_{min}	N_{max}	μ_N	σ_N
1,250	1,250	59,056	144,728	98,768	22,527
1,325	1,325	31,474	92,805	63,904	13,693
1,350	1,350	30,958	83,492	58,812	13,354
1,375	1,375	30,872	67,836	49,352	10,574
1,400	1,400	30,485	68,979	46,352	11,093

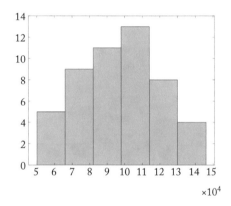

Figure 2.9: Histogram for the cyclic stress level: $\sigma_a = 22.083$ (ksi) and $\sigma_m = 22.083$ (ksi).

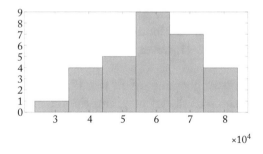

Figure 2.10: Histogram for the cyclic stress level: $\sigma_a = 22.5$ (ksi) and $\sigma_m = 22.5$ (ksi).

stress level $F_a = F_m = 1325$ (lb). Figure 2.10 is the histogram of N with a sample size 30 at the third stress level $F_a = F_m = 1350$ (lb). Figure 2.11 is the histogram of N with a sample size 30 at the fourth stress level $F_a = F_m = 1375$ (lb). Figure 2.12 is the histogram of N with a sample size 30 at the fifth stress level $F_a = F_m = 1400$ (lb).

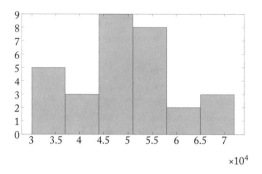

Figure 2.11: Histogram for the cyclic stress level: $\sigma_a = 22.917$ (ksi) and $\sigma_m = 22.917$ (ksi).

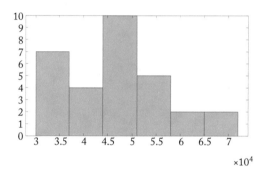

Figure 2.12: Histogram for the cyclic stress level: $\sigma_a = 23.333$ (ksi) and $\sigma_m = 23.333$ (ksi).

The histograms shown in Figures 2.8, 2.9, and 2.10 indicate that the number of cycles to failures N might follow a lognormal distribution. The histograms shown in Figures 2.11 and 2.12 do not suggest any distribution. However, they are not symmetric. So, they might also be a lognormal distribution.

We can use the Chi-Square goodness-of-fit test [7, 8, 14] to check whether they are a lognormal distribution. In the MATLAB program, the function "chi2gof" can be used to conduct the Chi-Square goodness-of-fit test, which has been discussed in Section 2.13 of the book [8]. The Chi-Square goodness-of-fit test on these data shows that the number of cycles to failures in each loading level can be treated as a lognormal distribution. The P-N curves for this set of fatigue test data with corresponding distribution parameters are listed in Table 2.13. In Table 2.13, $\mu_{\ln N}$ and $\sigma_{\ln N}$ are the log mean and log standard deviation of the lognormal distributed number of cycles to failures N.

Dr. E. B. Haugen in 1980 provided a set of P-S distributions of several materials under fully reversed bending loading or fully reversed axial loading [9]. Table 2.14 lists the P-S distributions at given fatigue life of three different materials. In Table 2.14, the fatigue strength S'_f has the unit of ksi. For three-parameter Weibull Distribution of S'_f, γ is the location parameter,

Table 2.13: The P-N distribution at different fatigue strength levels

Axial Stress Level #	σ_a (ksi)	σ_m (ksi)	Equivalent Stress σ_{a-eqi} (ksi)	Lognormal Distributed Fatigue Life	
				$\mu_{\ln N}$	$\sigma_{\ln N}$
1	20.833	20.833	35.126	11.4736	0.238106
2	22.083	22.083	38.832	11.0410	0.227320
3	22.5	22.5	40.139	10.9551	0.242377
4	22.917	22.917	41.485	10.7831	0.225735
5	23.333	23.333	42.871	10.716	0.241955

β is the shape parameter, and η is the scale parameters. For the normal distribution of S'_f, $\mu_{S'_f}$ is a mean and $\sigma_{S'_f}$ is a standard deviation. For steels, the P-S distribution at the fatigue life $N = 10^6$ (number of cycles to failure) in Table 2.14 is the distribution function of the material endurance limit.

2.8.2   THE COMPONENT P-S-N CURVES

There are two sets of distribution functions of a material P-S-N curves, which are the P-N distribution of the material fatigue life N at a given stress level and the P-S distribution of the material fatigue strength S'_f at a given fatigue life N. After the differences between the material fatigue test specimens and the component are considered by some modification factors, accordingly, we will have two sets of distribution functions of a component P-S-N curves, which will be the $P - N_c$ distribution of the component fatigue life N_c and the $P - S_f$ distribution of the component fatigue strength S_f.

For the component fatigue strength S_f at the given fatigue life N, we can provide two approaches to modify the material fatigue strength S'_f. In the first approach (the traditional approach), the surface finish modification factor k_a, the size modification factor k_b, and the loading modification factor k_c are used to modify the material fatigue strength S'_f to get the component fatigue strength S_f, as shown in Equation (2.13). While the fatigue strength concentration factor K_f is used to modify the stress amplitude σ_{a-eq}. The limit state function of a component at the given fatigue life N will be:

$$g\left(S_f, K_f, \sigma_{a-eq}\right) = S_f - K_f\sigma_{a-eq} = k_ak_bk_cS'_f - K_f\sigma_{a-eq} = \begin{cases} > 0 & \text{Safe} \\ 0 & \text{Limit state} \\ < 0 & \text{Failure.} \end{cases} \quad (2.27)$$

Table 2.14: The P-S distributions at different fatigue life of three materials

AISI 1045 Steel					
Rotary Bending, WQ for 1520°F Tempered at 1210°F, $K_t = 1$; $S_{ut} = 105$ ksi; $S_y = 82$ ksi					
Fatigue Life (number of cycles to failure)	3-Parameter Weibull Distribution of S_f'			Normal Distribution of S_f'	
	γ	β	η	$\mu_{S_f'}$	$\sigma_{S_f'}$
10^4	79.0	2.6	86.2	85.40	2.640
10^5	67.0	2.75	73.0	72.34	2.092
10^6	56.7	2.85	61.65	61.11	1.672
AISI 3140 Steel					
Rotary Bending, OQ for 1520°F Tempered at 1300°F, $K_t = 1$; $S_{ut} = 108$ ksi; $S_y = 87$ ksi					
Fatigue Life (number of cycles to failure)	3-Parameter Weibull Distribution			Normal Distribution	
	γ	β	η	$\mu_{S_f'}$	$\sigma_{S_f'}$
10^3	89	3.7	100.4	99.29	3.063
10^4	74	5.2	87.7	86.61	2.740
10^5	66	5	76.7	75.82	2.213
10^6	57	5.5	67.2	66.42	1.942
A-286 Stainless steel					
Fully Reversed Axial Load, $K_t = 1$; $S_{ut} = 90$ ksi; $S_y = 46$ ksi					
Fatigue Life (number of cycles to failure)	3-Parameter Weibull Distribution			Normal Distribution	
	γ	β	η	$\mu_{S_f'}$	$\sigma_{S_f'}$
10^4	40	1.84	54	52.44	7.072
10^5	31	2.1	43	41.63	5.347
10^6	24	2.2	34	32.856	4.267

For the reliability calculation, since K_f is always larger than 1, the above limit state function can be converted into the following equivalent limit state function,

$$g\left(S_f, K_f, \sigma_{a-eq}\right) = \frac{k_a k_b k_c}{K_f} S_f' - \sigma_{a-eq} = \begin{cases} > 0 & \text{Safe} \\ 0 & \text{Limit state} \\ < 0 & \text{Failure.} \end{cases} \qquad (2.28)$$

We will use the symbol S_{cf} to represent the component fatigue strength with the consideration of all modification factors including the fatigue stress concentration factor K_f:

$$S_{cf} = \frac{k_a k_b k_c}{K_f} S_f' \qquad \text{at the given fatigue life } N. \qquad (2.29)$$

Equation (2.29) can be used to build the limit state function Equation (2.28) of the component under cyclic stress at the given fatigue life. This result is the second approach to obtain component fatigue strength at the given fatigue life. We can assume that the modification factors will not change the type of distributions, but only change the distribution parameters. So, we can assume that S_{cf} will have the same type of distribution as that of S_f'. If S_f' follows a normal distribution with a mean $\mu_{S_f'}$ and a standard deviation $\sigma_{S_f'}$, the normally distributed S_{cf} will have the following mean $\mu_{S_{cf}}$ and standard deviation $\sigma_{S_{cf}}$:

$$\mu_{S_{cf}} = \frac{\mu_{k_a} k_b \mu_{k_c}}{\mu_{K_f}} \mu_{S_f'} \tag{2.30}$$

$$\sigma_{S_{cf}} = \frac{\mu_{k_a} k_b \mu_{k_c}}{\mu_{K_f}} \mu_{S_f'} \sqrt{\left(\frac{\sigma_{k_a}}{\mu_{k_a}}\right)^2 + \left(\frac{\sigma_{k_c}}{\mu_{k_c}}\right)^2 + \left(\frac{\sigma_{K_f}}{\mu_{K_f}}\right)^2 + \left(\frac{\sigma_{S_f'}}{\mu_{S_f'}}\right)^2}, \tag{2.31}$$

where μ_{k_a} and σ_{k_a} are the mean and the standard deviation of the surface finish modification factor k_a and determined per Equations (2.14), (2.15), and (2.16). k_b is the size modification factor and determined per Equation (2.17). μ_{k_c} and σ_{k_c} are the mean and the standard deviation of the load modification factor k_c and determined per Equations (2.18), (2.19), and (2.20). μ_{K_f} and σ_{K_f} are the mean and standard deviation of the fatigue stress concentration factor K_f and determined per Equations (2.22), (2.23), (2.24), and (2.25). $\mu_{S_f'}$ and $\sigma_{S_f'}$ are the mean and the standard deviation of material fatigue strength S_f' at the given fatigue life. $\mu_{S_{cf}}$ and $\sigma_{S_{cf}}$ are the mean and the standard deviation of the component fatigue strength at the given fatigue life.

For the component fatigue life N_c at the given cyclic stress level, we can use the following approach to modify the material fatigue life N. In a traditional S-N curve per Equation (2.1), the component S-N curve will be:

$$N \left(\frac{k_a k_b k_c}{K_f} S_f'\right)^m = constant. \tag{2.32}$$

We can rearrange Equation (2.32) as follows:

$$N \left(\frac{k_a k_b k_c}{K_f}\right)^m \left(S_f'\right)^m = constant. \tag{2.33}$$

Based on Equation (2.33), we can use N_c to represent the component fatigue life after the consideration of all modification factors at the given cyclic stress level:

$$N_c = N \left(\frac{k_a k_b k_c}{K_f}\right)^m. \tag{2.34}$$

It is assumed that the component fatigue life N_c has the same type of distribution as that of N. If N is a lognormal distribution with a log-mean $\mu_{\ln N}$ and log-standard deviation $\sigma_{\ln N}$, the distribution parameters of the component fatigue life N_c will be:

$$\mu_{\ln N_c} = \mu_{\ln N} + m \times \ln \left(\frac{\mu_{k_a} k_b \mu_{k_c}}{\mu_{K_f}} \right) \tag{2.35}$$

$$\sigma_{\ln N_c} = \sqrt{(\sigma_{\ln N})^2 + m^2 \left[\left(\frac{\sigma_{k_a}}{\mu_{k_a}} \right)^2 + \left(\frac{\sigma_{k_c}}{\mu_{k_c}} \right)^2 + \left(\frac{\sigma_{K_f}}{\mu_{K_f}} \right)^2 \right]}, \tag{2.36}$$

where $\mu_{\ln N}$ and $\sigma_{\ln N}$ are the log mean and the log standard deviation of the material fatigue life at the given cyclic stress level. $\mu_{\ln N_c}$ and $\sigma_{\ln N_c}$ are the log mean and the log standard deviation of the component fatigue life at the given cyclic stress level.

Figure 2.13 schematically depicts the material S-N curve per Equation (2.1) and the component S-N curve per Equation (2.32). The component fatigue life N_c per Equation (2.34) is schematically displayed by a horizontal line. At the given cyclic stress level S_f', which is exactly equal to the cyclic loading stress level, the material fatigue life is shrinking along the horizontal line from N to N_C. In the P-S-N curve approach, N_C is a random variable and will be described by a distribution function.

The component fatigue strength S_{cf} per Equation (2.29) is schematically displayed by a vertical line in Figure 2.13. At the given fatigue life N, which is exactly equal to the number of cycles of the cyclic loading stress, the material fatigue strength S_f' is shrinking along the vertical line from S_f' to S_{Cf}. In the P-S-N curve approach, S_{Cf} is a random variable and will be described by a distribution function.

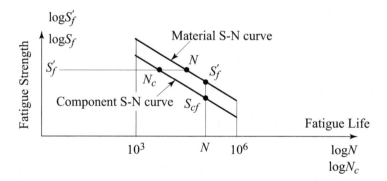

Figure 2.13: Schematic of a material S-N curve and a component S-N curve.

The applications of the $P - N_C$ curves at a given cyclic loading stress level or $P - S_{Cf}$ curves at a given fatigue life will be demonstrated in the following sections.

2.8.3   RELIABILITY OF A COMPONENT UNDER MODEL #1 CYCLIC LOADING SPECTRUM

The general description of model #1 is $(\sigma_a, \sigma_m, n_L)$, where σ_a is a constant stress amplitude of the cyclic stress, σ_m is constant mean stress of the cyclic stress, and n_L is the number of cycles of the cyclic loading. Since provided P-S-N curves are typically obtained based on fully reversed cyclic stress fatigue tests, the non-zero-mean cyclic stress will be converted into an equivalent stress amplitude σ_{a-eq} of a fully reversed cyclic stress per Equation (2.21).

When the $P - S_{cf}$ curve of component fatigue strength S_{cf} at the given fatigue life $N = n_L$ are provided, the reliability of the component can be directly calculated by the following equation:

$$R = P\left(S_{cf} > \sigma_{a-eq}\right) = 1 - \int_{-\infty}^{\sigma_{a-eq}} f_{S_{cf}}(s)ds = 1 - F_{S_{cf}}\left(\sigma_{a-eq}\right), \qquad (2.37)$$

where $f_{S_{cf}}(s)$ and $F_{S_{cf}}(s)$ are the probability density function (PDF) and cumulative distribution function (CDF) of the component fatigue strength S_{cf} at the given fatigue life N which is equal to the number of cycles n_L of the model #1 cyclic stress. σ_{a-eq} is a constant equivalent stress amplitude of the cyclic stress.

When the $P - S_{cf}$ curve of component fatigue strength S_{cf} at the given fatigue life $N = n_L$ are obtained through the material P-S-N curves, that is per Equation (2.29), the component fatigue strength at the fatigue life $N = n_L$ is:

$$S_{cf} = \frac{k_a k_b k_c}{K_f} S'_f \qquad \text{at the given fatigue life } N. \qquad (2.29)$$

Then the limit state function of the component under this situation is:

$$g\left(k_a, k_c, K_f, S'_f\right) = \frac{k_a k_b k_c}{K_f} S'_f - \sigma_{a-eq} = \begin{cases} > 0 & \text{Safe} \\ 0 & \text{Limit state} \\ < 0 & \text{Failure,} \end{cases} \qquad (2.38)$$

where k_a, k_b, and k_c are the surface finish modification factor, the size modification factor, and the loading modification factor, respectively. K_f is the fatigue stress concentration factor. S'_f is the material fatigue strength at the fatigue life $N = n_L$. σ_{a-eq} is a constant equivalent stress amplitude of the cyclic stress, which can be calculated per Equation (2.21). In Equation (2.38), k_b and σ_{a-eq} will be treated as deterministic constants. The reliability of the component under such a cyclic loading can be calculated by using the limit state function Equation (2.38) with the H-L method, R-F method, or Monte Carlo method.

When the $P - N_c$ curves of component fatigue life N_c at the given fatigue cyclic stress level $S'_f = \sigma_{a-eq}$ are provided, the reliability of the component can be directly calculated by the following equation:

$$R = P\left(N_c > n_L\right) = 1 - \int_{-\infty}^{n_L} f_{N_c}\left(n\right) dn = 1 - F_{N_c}\left(n_L\right), \tag{2.39}$$

where $f_{N_c}\left(n\right)$ and $F_{N_c}\left(n\right)$ are the PDF and CDF of the component fatigue life N_c at the given cyclic stress level S_f' which is equal to σ_{a-eq} of model #1 cyclic stress. n_L is the number of cycles of the model #1 cyclic loading stress. n_L is a constant number for the model #1 cyclic loading stress

When the $P - N_c$ curve of component fatigue life N_c at the given fatigue cyclic stress level $S_f' = \sigma_{a-eq}$ are obtained through the material P-S-N curves, that is per Equation (2.34), the limit state function of this problem is:

$$g\left(k_a, k_c,\ K_f, N\right) = N\left(\frac{k_a k_b k_c}{K_f}\right)^m - n_L = \begin{cases} > 0 & \text{Safe} \\ 0 & \text{Limit state} \\ < 0 & \text{Failure,} \end{cases} \tag{2.40}$$

where k_a, k_b, k_c, and K_f are the same as those in Equation (2.38). N is the material fatigue life at the cyclic stress level $S_f' = \sigma_{a-eq}$. n_L is a constant number for the model #1 cyclic loading stress. The reliability of the component under such a cyclic loading can be calculated by using the limit state function Equation (2.40) with the H-L method, R-F method, or Monte Carlo method.

For a component under model #1 cyclic loading spectrum, we could use Equations (2.37) or (2.39) to directly calculate component's reliability. Or, we can use the limit station functions (2.38) and (2.40) to calculate component's reliability. Three examples are presented to show how to calculate the reliability of a component under model #1 cyclic loading spectrum.

Example 2.9
A beam at its critical section is subjected to a fully reversed cyclic bending stress with a constant stress amplitude $\sigma_a = 35.5$ (ksi) and a constant number of cycles $n_L = 3.90 \times 10^5$ (cycles). After a series of calculation, the component fatigue life of the beam at its critical section under the fully reversed cyclic bending stress $S_{cf} = 35.5$ (ksi) follows a lognormal distribution with a log mean $\mu_{\ln N} = 13.305$ and a log standard deviation $\sigma_{\ln N} = 0.187$. Calculate the reliability of the beam.

Solution:
For this problem, the component fatigue life $P - N_C$ distribution at the given cyclic stress level is given. We can directly use Equation (2.39) to calculate the reliability of the beam.

If the Microsoft Excel formula is used,

$$R = P\left(N_c > n_L\right) = 1 - F_{N_c}\left(n_L\right)$$
$$= 1 - LOGNORM.DIST\left(3.90 \times 10^5,\ 13.305, 0.187,\ true\right) = 0.9894.$$

If the MATLAB function is used,

$$R = P\,(N_c > n_L) = 1 - F_{N_c}\,(n_L) = 1 - logncdf\,(3.90 \times 10^5,\ 13.305, 0.187) = 0.9894.$$

■

Example 2.10

A constant cross-section bar is subject to a fully reversed cyclic axial stress with a constant stress amplitude $\sigma_a = 23.5$ (ksi) and a constant number of cycles $n_L = 1 \times 10^4$ (cycles). The material fatigue life at the fully reversed rotating bending stress amplitude $S'_f = 23.5$ (ksi) follows a lognormal distribution with a log mean $\mu_{\ln N} = 13.72$ and a log standard deviation $\sigma_{\ln N} = 0.124$. The bar can be treated as a hot-rolled component. Its ultimate strength is 45.4 (ksi), and its slope of the traditional S-N curve m is 8.30. (1) Determine the distribution parameters of the component fatigue life at the given cyclic stress level. (2) Calculate the reliability of this bar.

Solution:

(1) Determine the distribution parameters of the component fatigue life N_c at the given cyclic stress level.

Since the material fatigue life distribution at the given cyclic stress level is provided, Equation (2.34) will be used to get the component fatigue life.

Per Equations (2.14)–(2.16), for a hot-rolled component, the distribution parameters of the surface finish modification factor k_a are:

$$\mu_{k_a} = 16.45\,(S_{ut})^{-0.7427} = 16.45\,(45.4)^{-0.7427} = 0.9671 \qquad \text{(a)}$$

$$\sigma_{k_a} = \gamma_{k_a} \times \mu_{k_a} = 0.098 \times 0.9671 = 0.09478. \qquad \text{(b)}$$

Per Equation (2.17), the size modification factor k_b is:

$$k_b = 1. \qquad \text{(c)}$$

Per Equations (2.18)–(2.20), the loading modification factor k_c for cyclic axial loading are:

$$\mu_{k_c} = 0.774. \qquad \text{(d)}$$

$$\sigma_{k_c} = \gamma_{k_c} \times \mu_{k_c} = 0.163 \times 0.774 = 0.1262. \qquad \text{(e)}$$

In this example, K_f will be 1 because of a constant cross-section bar.

The component fatigue life at the given cyclic stress level will be:

$$N_c = N\,(k_a k_c)^{8.30}. \qquad \text{(f)}$$

If it is assumed that the component fatigue life N_c still follows a log-normal distribution, we can calculate its log-mean and log-standard deviation per Equations (2.35) and (2.36):

$$\mu_{\ln N_c} = \mu_{\ln N} + m \times \ln\left(\mu_{k_a}\mu_{k_c}\right)$$
$$= 13.72 + 8.30 \times \ln\left(0.9671 \times 0.774\right) = 11.393 \tag{g}$$

$$\sigma_{\ln N_c} = \sqrt{(\sigma_{\ln N})^2 + m^2\left[\left(\frac{\sigma_{k_a}}{\mu_{k_a}}\right)^2 + \left(\frac{\sigma_{k_c}}{\mu_{k_c}}\right)^2\right]}$$
$$= \sqrt{(0.124)^2 + 8.30^2\left[\left(\frac{0.09478}{0.9671}\right)^2 + \left(\frac{0.1262}{0.774}\right)^2\right]} = 1.5838 \tag{h}$$

(2) Calculate the reliability of this bar.

If we use the distribution parameters in Equations (g) and (h), that is, N_c is a log-normal distribution with $\mu_{\ln N_c} = 11.393$ and $\sigma_{\ln N_c} = 1.5838$, the reliability of the bar per Equation (2.40) will be:

$$R = P\left(N_c > n_L\right) = 1 - F_{N_c}\left(n_L\right) = 1 - logncdf\left(1000, 11.393, 1.5838\right) = 0.9159.$$

Of course, this is only an approximate result. If the limit state function Equation (2.40) for this example is used, we need to use the R-F or Monte Carlo method to compile a MATLAB program to calculate the reliability of the bar and will get a more accurate result.

∎

Example 2.11

The critical section of a machined rotating shaft is at its shoulder section. The schematic of the shoulder section is shown in Figure 2.14. The critical section is subjected to fully reversed cyclic bending stress with a constant stress amplitude $\sigma_a = 10.67$ (ksi) and a constant number of cycles $n_L = 3 \times 10^5$ (cycles). The material fatigue strength S_f' at the fatigue life $N = 3.5 \times 10^5$ from fully reversed rotating bending stress tests follows a normal distribution with a mean $\mu_{S_f'} = 26.52$ (ksi) and standard deviation $\sigma_{S_f'} = 1.98$ (ksi). The ultimate strength of this material is 61.5 (ksi). (1) Express the component fatigue strength at the fatigue life $N = 3.5 \times 10^5$. (2) Calculate the reliability of the shaft.

Solution:

(1) The component fatigue strength at the fatigue life $N = 3.5 \times 10^5$:

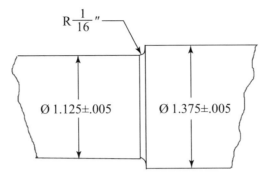

Figure 2.14: Schematic of the shoulder section of a shaft.

Per Equation (2.29), we can have the expression of the component fatigue strength S_{cf} at the fatigue life $N = 3.5 \times 10^5$:

$$S_{cf} = \frac{k_a k_b k_c}{K_f} S_f' \quad \text{at} \quad N = 3.5 \times 10^5. \tag{a}$$

Per Equations (2.14)–(2.16), the distribution parameters of the surface finish modification factor k_a for a machined component are:

$$\mu_{k_a} = 2.7 \, (S_{ut})^{-0.2653} = 2.7 \, (61.5)^{-0.2653} = 0.9053 \tag{b}$$

$$\sigma_{k_a} = \gamma_{k_a} \times \mu_{k_a} = 0.06 \times 0.9053 = 0.05432. \tag{c}$$

Per Equation (2.17), the size modification factor is:

$$k_b = \left(\frac{d}{0.3}\right)^{-0.1133} = \left(\frac{1.125}{0.3}\right)^{-0.1133} = 0.8609. \tag{d}$$

Per Equations (2.18)–(2.20), the distribution parameters of the loading modification factor k_c for cyclic bending stress, we have:

$$\mu_{k_c} = 1 \tag{e}$$

$$\sigma_{k_c} = 0. \tag{f}$$

Per Equations (2.22)–(2.25), we can calculate the mean and standard deviation of the fatigue stress concentration factor K_f. In this example, $K_t = 2.01$ and $r = 0.0625''$.

Per Equation (2.23), the Neuber constant $\sqrt{a}$ is

$$\sqrt{a} = \frac{4}{S_{ut}} = \frac{4}{61.5} = 0.06504. \tag{g}$$

Per Equation (2.22), the mean of K_f is

$$\mu_{K_f} = \frac{K_t}{1 + \frac{2}{\sqrt{r}}\left(\frac{K_t - 1}{K_t}\right)\sqrt{a}} = \frac{2.01}{1 + \frac{2}{\sqrt{0.0625}}\left(\frac{2.01 - 1}{2.01}\right)0.06504} = 1.5934. \qquad \text{(h)}$$

Per Equations (2.24) and (2.25), the standard deviation of K_f is:

$$\sigma_{K_f} = \mu_{K_f} \times \gamma_{K_f} = 1.5934 \times 0.08 = 0.1275. \qquad \text{(i)}$$

So, the component fatigue strength S_{cf} for this example is

$$S_{cf} = 0.8609 \frac{k_a}{K_f} S_f' \qquad \text{at} \quad N = 3.5 \times 10^5. \qquad \text{(j)}$$

The distribution parameters of every random variable in Equation (j) are all known, the component fatigue strength S_{cf} is fully specified.

(2) Calculate the reliability of the shaft.

The limit state function in this example per Equation (2.38) will be:

$$g\left(k_a, \, K_f, S_f'\right) = 0.8609 \frac{k_a}{K_f} S_f' - 10.67. \qquad \text{(k)}$$

The distribution parameters in the limit state function (k) are listed in Table 2.15.

The limit state function (k) contains three normally distributed random variable and is a nonlinear function. We can follow the H-L method and the program flowchart in Appendix A.1 to create a MATLAB program. The iterative results are listed in Table 2.16. From the iterative results, the reliability index β and corresponding reliability R of the shaft in this example are:

$$\beta = 1.56120 \qquad R = \Phi(1.5620) = 0.9409.$$

∎

Table 2.15: The distribution parameters of random variables in Equation (k)

k_a		K_f		S_f' (ksi)	
μ_{k_a}	σ_{k_a}	μ_{K_f}	σ_{K_f}	$\mu_{S_f'}$	$\sigma_{S_f'}$
0.9053	0.05432	1.5932	0.1275	26.52	1.98

Table 2.16: The iterative results of Example 2.11 by the H-L method

Iterative #	$k_a^{'*}$	k_f^{*}	$S_f^{'*}$	β^{*}	$|\Delta\beta^{*}|$
1	0.9053	1.5932	21.8117	1.508306	
2	0.87095	1.718111	24.4495	1.559894	0.051589
3	0.865776	1.721417	24.64291	1.561837	0.001942
4	0.865396	1.721634	24.65684	1.561987	0.00015
5	0.865369	1.72165	24.65787	1.561999	1.18E-05

2.8.4   RELIABILITY OF A COMPONENT UNDER MODEL #2 CYCLIC LOADING SPECTRUM

Model #2 cyclic loading is one constant cyclic stress level σ_a with a distributed number of cycles n_L. When a component is subjected to model #2 cyclic loading and if the component fatigue life N_C at the cyclic stress level σ_a is provided, we will have the following limit state function:

$$g\,(N_c, n_L) = N_c - n_L = \begin{cases} > 0 & \text{Safe} \\ 0 & \text{Limit state} \\ < 0 & \text{Failure,} \end{cases} \qquad (2.41)$$

where N_c is the component fatigue life at the fatigue strength $S_f' = \sigma_a$.

If the component fatigue life N_C at the cyclic loading stress loading level σ_a is obtained per Equation (2.34), the limit state function of a component under the model #2 cyclic loading spectrum will be:

$$g\,\left(k_a, k_c, K_f, N, n_L\right) = N\left(\frac{k_a k_b k_c}{K_f}\right)^m - n_L = \begin{cases} > 0 & \text{Safe} \\ 0 & \text{Limit state} \\ < 0 & \text{Failure,} \end{cases} \qquad (2.42)$$

where k_a, k_b, k_c, and K_f are the surface finish modification factor, the size modification factor, the loading modification factor, and the fatigue stress concentration factor, respectively. k_a, k_c, and K_f will be normally distributed random variables. k_b will be treated as a deterministic constant. N is a distributed material fatigue life at the fatigue strength $S_f' = \sigma_a$. n_L is a distributed number of cycles of the cyclic stress loading σ_a.

Both limit state function (2.41) and (2.42) can be used to calculate the reliability of a component under model #2 cyclic loading spectrum by the H-L, R-F, or Monte Carlo method. We will use the limit state function (2.41) to demonstrate an example.

Example 2.12

A component is subjected to a fully reversed cyclic bending stress with a constant stress amplitude $\sigma_a = 26.6$ (ksi). The number of cycles n_L of this cyclic loading can be described by a normal distribution with a mean $\mu_{n_L} = 4.25 \times 10^4$ (cycles) and a standard deviation $\sigma_{n_L} = 3253$ (cycles). The fatigue life of this component N_c at the fatigue strength $S'_f = \sigma_a = 26.6$ (ksi) follows a lognormal distribution with a log-mean $\mu_{\ln N_c} = 11.01$ and the standard deviation $\sigma_{\ln N_c} = 0.158$. Calculate the reliability of the component.

Solution:

Since the component fatigue life N_c of this component at the cyclic stress level $\sigma_a = 26.6$ (ksi) is given, we can use Equation (2.41) to build the limit state function:

$$g\left(N_c, n_L\right) = N_c - n_L = \begin{cases} > 0 & \text{Safe} \\ 0 & \text{Limit state} \\ < 0 & \text{Failure.} \end{cases} \tag{a}$$

The distribution parameters of the limit state function (a) are listed in Table 2.17.

The limit state function (a) contains one normally distributed random variable and one lognormal distribution. We can use the R-F method to calculate its reliability, which is presented in Appendix A.2, to compile a MATLAB program for this example. The iterative results are listed in Table 2.18. From the iterative results, the reliability index β and corresponding reliability R of the component in this example are:

$$\beta = 2.02131 \qquad R = \Phi\left(2.02131\right) = 0.9784.$$

∎

Table 2.17: The distribution parameters of random variables in Equation (a)

N_C (lognormal distribution)		n_L (normal distribution)	
$\mu_{\ln N_C}$	$\sigma_{\ln N_C}$	μ_{n_L}	σ_{n_L}
11.01	0.158	42500	3253

2.8.5   RELIABILITY OF A COMPONENT UNDER MODEL #3 CYCLIC LOADING SPECTRUM

Model #3 cyclic loading spectrum is one constant number of cycles n_L with a distributed cyclic stress amplitude σ_a. When a component is subjected to model #3 cyclic loading and if the component fatigue strength S_{Cf} at the fatigue life $N = n_L$ is provided, we will have the following

Table 2.18: The iterative results of Example 2.12 by the R-F method

Iterative #	N_c^*	n_L^*	β^*	$|\Delta\beta^*|$
1	61,235.48	61,235.48	1.760759	
2	44,322.18	44,322.18	2.02009	0.259331
3	45,263.78	45,263.78	2.021308	0.001218
4	45,217.85	45,217.85	2.021311	2.82E-06

limit state function:

$$g\left(S_{Cf}, \sigma_a\right) = S_{Cf} - \sigma_a = \begin{cases} > 0 & \text{Safe} \\ 0 & \text{Limit state} \\ < 0 & \text{Failure,} \end{cases} \quad (2.43)$$

where S_{Cf} is the component fatigue strength at the fatigue life N, which is equal to the number of cycles n_L of model #3 cyclic loading spectrum.

If the component fatigue strength S_{Cf} at the fatigue life $N = n_L$ is obtained per Equation (2.39), the limit state function of a component under the model #3 cyclic loading spectrum will be:

$$g\left(k_a, k_c, K_f, S_f', \sigma_a\right) = S_{Cf} - \sigma_a = \frac{k_a k_b k_c}{K_f} S_f' - \sigma_a = \begin{cases} > 0 & \text{Safe} \\ 0 & \text{Limit state} \\ < 0 & \text{Failure,} \end{cases} \quad (2.44)$$

where k_a, k_b, k_c, and K_f are the surface finish modification factor, the size modification factor, the loading modification factor, and the fatigue stress concentration factor, respectively. k_a, k_c, and K_f will be normally distributed random variables. k_b will be treated as a deterministic constant. S_f' is a distributed material fatigue strength at the fatigue life $N = n_L$. σ_a is a distributed fully reversed cyclic stress amplitude of the model #3 cyclic loading with the given constant number of cycles n_L.

Both limit state functions (2.43) and (2.44) can be used to calculate the reliability of a component under model #3 cyclic loading with the H-L, R-F, or Monte Carlo method. We will use Equation (2.44) to run one example.

Example 2.13
The critical section of a machined stepped plate is at its stepped section, as shown in Figure 2.15. The plate has a thickness $t = 0.375 \pm 0.010''$. The plate is subjected to a fully reversed axial loading amplitude F_a which is a normal distribution with a mean $\mu_{F_a} = 5.2$ (klb) and a standard deviation $\sigma_{F_a} = 0.61$ (klb). The number of cycles of this fully reversed axial loading is $n_L =$

3×10^5 (cycles). The ultimate strength of this material is 61.5 (ksi). The material fatigue strength S'_f at the fatigue life $N = 3.5 \times 10^5$ from fully reversed rotating bending stress tests follows a normal distribution with a mean $\mu_{S'_f} = 26.52$ (ksi) and standard deviation $\sigma_{S'_f} = 1.98$ (ksi). Use the Monte Carlo method to calculate the reliability of this plate and its range with a 95% confidence level.

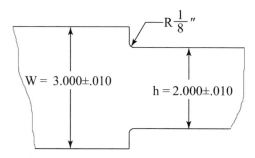

Figure 2.15: Schematic of the stepped section of a plate.

Solution:

(1) Build the limit state function at the stepped section of the plate.

We will use Equation (2.44) to build the limit state function at the stepped section of the plate. Per Equation (2.39), we can have the expression of the component fatigue strength S_{cf} at the fatigue life $N = 3.5 \times 10^5$:

$$S_{cf} = \frac{k_a k_b k_c}{K_f} S'_f \qquad \text{at} \quad N = 3.5 \times 10^5. \tag{a}$$

Per Equations (2.14)–(2.16), for a machined component, the mean and standard deviation of the surface finish modification factor k_a are:

$$\mu_{k_a} = 0.9053; \qquad \sigma_{k_a} = 0.05432. \tag{b}$$

Per Equation (2.17), the size modification factor k_b for cyclic axial loading is

$$k_b = 1. \tag{c}$$

Per Equations (2.18)–(2.20), for cyclic axial loading, the mean and the standard deviation of the loading modification factor are:

$$\mu_{k_c} = 0.774; \qquad \sigma_{k_c} = 0.1262. \tag{d}$$

In this example, $K_t = 2.14$ and $r = 0.25''$. Per Equations (2.22)–(2.25), we can calculate the mean and standard deviation of the fatigue stress concentration factor K_f:

$$\mu_{K_f} = 1.880; \qquad \sigma_{K_f} = 0.1504. \tag{e}$$

So, the component fatigue strength S_{cf} for this example is

$$S_{cf} = \frac{k_a}{K_f} S_f' \quad \text{at} \quad N = 3.5 \times 10^5. \tag{f}$$

The fully reversed stress amplitude σ_a due to the fully reversed axial loading F_a in this example is:

$$\sigma_a = \frac{F_a}{h \times t}. \tag{g}$$

The mean and standard deviation of geometric dimensions h per Equation (1.1) are:

$$\mu_h = 2'', \qquad \sigma_h = 0.0025''. \tag{h}$$

The mean and standard deviation of geometric dimensions t per Equation (1.1) are:

$$\mu_t = 0.375'', \qquad \sigma_h = 0.0025''. \tag{i}$$

So, the limit state function at the stepped section of this plate per Equation (2.44) is

$$g\left(k_a, k_c, K_f, S_f', h, t, F_a\right) = S_{Cf} - \sigma_a = \frac{k_a k_c}{K_f} S_f' - \frac{F_a}{h \times t} = \begin{cases} > 0 & \text{Safe} \\ 0 & \text{Limit state} \\ < 0 & \text{Failure.} \end{cases} \tag{j}$$

The distribution parameters of every random variable in Equation (j) are all normal distributions. Their distribution parameters are listed in Table 2.19.

Table 2.19: The distribution parameters of random variables for Example 2.13 in Equation (j)

k_a		k_c		K_f		S_f' (ksi)		h		t		F_a (klb)	
μ_{k_a}	σ_{k_a}	μ_{k_c}	σ_{k_c}	μ_{K_f}	σ_{K_f}	$\mu_{S_f'}$	$\sigma_{S_f'}$	μ_h	σ_h	μ_t	σ_t	μ_{F_a}	σ_{F_a}
0.9503	0.05432	0.774	0.1262	1.880	0.1504	26.52	1.98	2	0.0025	0.375	0.0025	5.2	0.61

(2) Reliability of the stepped plate and its range with a 95% confidence level.

We can use the Monte Carlo simulation method to calculate the reliability of this example. We can follow the Monte Carlo method and the program flowchart in Appendix A.3 to create a MATLAB program. The estimated reliability of this component R, the probability of failure F, and the relative error ε of F are:

$$R = 0.9464, \qquad F = 0.0536, \qquad \varepsilon = 0.0066.$$

So, the range of the probability of failure with a 95% confidence level will be:

$$F = 0.0536 \pm 0.0536 \times 0.0066 = 0.0536 \pm 0.0004.$$

Therefore, the range of the reliability of the component with a 95% confidence level will be:

$$R = 1 - F = 0.9464 \pm 0.0004.$$

∎

2.8.6   RELIABILITY OF A COMPONENT UNDER MODEL #4 CYCLIC LOADING SPECTRUM

Model #4 cyclic loading spectrum is multiple constant cyclic stress levels with multiple constant numbers of cycles, which has been discussed in Section 1.2. When a component is under the model #4 cyclic loading spectrum, no direct limit state function can be established. However, the equivalent fatigue damage concept [11, 15] proposed by Dr. Dimitri B. Kececioglu in 1977 can be used to estimate the reliability of the component under such cyclic loading spectrum. The assumption in the approach is that the cyclic number at a cyclic stress level could be transferred to another stress level with an equivalent cyclic number under the condition that the probability of a safe status of the component at the original stress level is the same as that at the transferred stress level with the equivalent cyclic number. In this approach, the reliability index of the component under cyclic loading is used as an indirect index for measuring the fatigue damage of a component.

Let us use two levels of the model #4 cyclic loading spectrum as listed in Table 2.20 to demonstrate the equivalent fatigue damage concept and procedure. The corresponding fatigue life distributions are also listed in Table 2.20. In Table 2.20, σ_{ai} and σ_{aj} are the fully reversed cyclic stress amplitudes in the cyclic stress levels i and j. n_{Li} and n_{Lj} are the numbers of cycles in the cyclic stress levels i and j. They are all constants for the model #4 cyclic loading spectrum, that is, deterministic values. S_f' is material fatigue strength, which is equal to the cyclic stress amplitude of the corresponding stress level, as shown in Table 2.20. The component fatigue life N_C at the given fatigue strength $S_f' = \sigma_{ai}$ is typically a lognormal distribution, which has been discussed in Section 2.8.2. It can be obtained per Equation (2.34). Its distribution parameters can be calculated per Equations (2.35) and (2.36). In Table 2.20, $\mu_{\ln N_{Ci}}$ and $\sigma_{\ln N_{Ci}}$ are the mean and the standard deviation of the component fatigue life N_{Ci} at the cyclic stress level i. $\mu_{\ln N_{Cj}}$ and $\sigma_{\ln N_{Cj}}$ are the mean and the standard deviation of the component fatigue life N_{Cj} at the cyclic stress level j.

The general procedure for transferring cyclic stress level from stress level i to the stress level j is described as the following.

Step 1: Calculate the index of the fatigue damage of the component due to the cyclic stress (σ_{ai}, n_{Li}) in the stress level i.

The index of the fatigue damage of the component due to the cyclic stress (σ_{ai}, n_{Li}) in the stress level #i can be indirectly represented by the probability $P(N_C > n_{Li}) = P[\ln(N_C) > \ln(n_{Li})]$,

Table 2.20: Two cyclic stress levels and corresponding component fatigue life

Model #4 Cyclic Loading Spectrum			Component Fatigue Life N_C (lognormal distribution)			
Stress level #	Cyclic stress	Number of cycles	Cyclic stress	$\mu_{\ln N_C}$	$\sigma_{\ln N_C}$
i	σ_{ai}	n_{Li}	$S_f' = \sigma_{ai}$	$\mu_{\ln N_{Ci}}$	$\sigma_{\ln N_{Ci}}$
j	σ_{aj}	n_{Lj}	$S_f' = \sigma_{aj}$	$\mu_{\ln N_{Cj}}$	$\sigma_{\ln N_{Cj}}$

which is the reliability of the component under the cyclic stress level #i. The reliability can be directly represented by the reliability index β. Therefore, we can use the reliability index to represent fatigue damage due to the cyclic loading. The reliability index β_i of the component in the cyclic stress level #i will be:

$$\beta_i = \frac{\mu_{\ln N_{Ci}} - \ln(n_{Li})}{\sigma_{\ln N_{Ci}}}. \tag{2.45}$$

Step 2: Calculate the equivalent number of cycles from the cyclic stress in stress level #i to the cyclic stress level #j.

Let n_{eqi-j} represent the equivalent number of cycles to the cyclic stress level #j. According to the equivalent fatigue damage concept, the reliability index of this n_{eqi-j} in the stress level #j will be β_i . Therefore, we have:

$$\beta_i = \frac{\mu_{\ln N_{Cj}} - \ln(n_{eqi-j})}{\sigma_{\ln N_{Cj}}}. \tag{2.46}$$

By rearranging Equation (2.46), we have:

$$n_{eqi-j} = \exp\left(\mu_{\ln N_{Cj}} - \beta_i \times \sigma_{\ln N_{Cj}}\right). \tag{2.47}$$

Step 3: Combine cyclic stresses of the two-stress levels.

The component fatigue damage due to the two cyclic stresses (σ_{ai}, n_{Li}) and (σ_{aj}, n_{Lj}) in two stress levels #i and #j will be equal to the component fatigue damage due to cyclic stress $(\sigma_{aj}, n_{Lj} + n_{eqi-j})$ in the stress level #j. So, the total equivalent number of cycles n_{j-eq} in the stress level j is:

$$n_{j-eq} = n_{Lj} + n_{eqi-j}, \tag{2.48}$$

where n_{j-eq} is the equivalent number of cycles, including the original number of cycles in current stress level and the transferred equivalent number of cycles from another stress level.

Step 4: Calculate the reliability of the component due to model #4 cyclic loading spectrum.

If there are more than two stress levels in the model #4 cyclic loading spectrum, we can repeat Step 1 to Step 3 to combine two stress levels into the next stress level, until we reach the last stress level. Let us assume that the j stress level is the last stress level of model #4 cyclic loading spectrum. The reliability of the component due to all stress levels will be:

$$R = \Phi \left[\frac{\mu_{\ln N_{Cj}} - \ln \left(n_{Lj} + n_{eqi-j} \right)}{\sigma_{\ln N_{Cj}}} \right]. \tag{2.49}$$

If the component fatigue life N_C follows another type of distribution, we can follow the above steps to run a similar calculation by using the equivalent fatigue damage concept.

Example 2.14

A component is subjected to model #4 cyclic loading spectrum, as shown in Table 2.21. The component fatigue life N_c under the corresponding stress levels are lognormal distributions, as shown in Table 2.21. Use the equivalent fatigue damage concept to calculate the reliability of the component.

Table 2.21: Model #4 cyclic loading spectrum for Example 2.14

Model #4 Cyclic Loading Spectrum			Component Fatigue Life N_C (lognormal distribution)		
Level #	Cyclic stress (psi)	Number of cycles	Cyclic stress (psi)	$\mu_{\ln N_C}$	$\sigma_{\ln N_C}$
1	65,000	81,000	65,000	12.95987	0.198
2	85,000	16,000	85,000	11.01311	0.197
3	105,000	2,800	105,000	9.47966	0.195

Solution:

We can use Equations (2.45), (2.47), and (2.48) twice to convert the number of cycles in stress level 1 to level 2, and then from stress level 2 to the last stress level 3.

For the stress level 1 to stress level 2, we have:

$$\beta_1 = \frac{\mu_{\ln N_{C1}} - \ln (n_{L1})}{\sigma_{\ln N_{C1}}} = \frac{12.95987 - \ln(81000)}{0.198} = 8.437205 \tag{a}$$

$$n_{eq1-2} = \exp \left(11.01311 - 8.437205 \times 0.197 \right) = 11658.8 \tag{b}$$

$$n_{2-eq} = n_{L2} + n_{eq1-2} = 16000 + 11658.8 = 27658.8. \tag{c}$$

For the stress level 2 to stress level 3, we have:

$$\beta_2 = \frac{\mu_{\ln N_{C2}} - \ln\left(n_{L2-eq}\right)}{\sigma_{\ln N_{C2}}} = \frac{11.01311 - \ln(27658.8)}{0.191} = 3.986858. \tag{d}$$

$$n_{eq2-3} = \exp\left(9.47966 - 3.986858 \times 0.195\right) = 6016.27 \tag{e}$$

$$n_{3-eq} = n_{L3} + n_{eq2-3} = 2800 + 6016.27 = 8816.27. \tag{f}$$

The stress level 3 is the last. Therefore, the reliability of the component in this example per Equation (2.49) is

$$R = \Phi\left[\frac{\mu_{\ln N_{C3}} - \ln\left(n_{3-eq}\right)}{\sigma_{\ln N_{C3}}}\right] = \Phi\left[\frac{9.47966 - \ln(8816.27)}{0.195}\right] = \Phi(2.02721) = 0.9787. \tag{g}$$

∎

In Example 2.14, if we convert the cyclic stress from the stress level 3 to the stress level 2, and then from the stress level 2 to the stress level 1, we will get the reliability $R = \Phi(2.07511) = 0.9810$. The results are slightly different because it is an approximate estimation with the assumption of the equivalent fatigue damage concept.

2.8.7 RELIABILITY OF A COMPONENT UNDER MODEL #5 CYCLIC LOADING SPECTRUM

Model #5 cyclic loading spectrum consists of multiple constant stress amplitudes of cyclic loadings with corresponding distributed cycle numbers at each cyclic stress level. This section will discuss how to calculate the reliability of a component under model #5 cyclic loading spectrum.

It is very difficult to create the limit state function of a component under model #5 cyclic loading spectrum. The author, in 2016, proposed an approach with a modified equivalent fatigue damage concept to deal with this type of problem [16]. Let us use two stress loading levels with a distributed number of cycles as an example to explain this approach. The cyclic loading and corresponding component fatigue life at two different stress levels are listed in Table 2.22. n_{Li} is a distributed number of cycles of the fully reversed cyclic loading with a stress amplitude σ_{ai} in the stress level #i. N_{Ci} is the distributed component fatigue life at the fully reversed fatigue strength $S_f' = \sigma_{ai}$ in the stress level #i. The component fatigue life N_{Ci} can be determined by Equations (2.34), (2.35), and (2.36), which have been discussed in Section 2.8.2. Two assumptions [16] for this approach are as follows.

Assumption One: The reliability index of the component under cyclic loading is used as an indirect index for measuring fatigue damage of a component. To transfer a distributed cyclic number n_{Li} at the stress level σ_{ai} to the distributed cyclic number n_{Lj} at the stress level σ_{aj}, the reliability index of the component due to n_{Li} at the cyclic stress level σ_{ai} should be equal to

Table 2.22: Two cyclic stress levels of model #5 cyclic loading spectrum

Stress Level #	Model #5 Cyclic Loading Spectrum		Component Fatigue Life N_C at the Given Fatigue Strength S'_f	
	Cyclic Stress Amplitude (constant)	Distributed Number of Cycles	Fatigue Strength (constant)	Distributed Fatigue Life N_C
i	σ_{ai}	n_{Li}	$S'_f = \sigma_{ai}$	N_{Ci}
j	σ_{aj}	n_{Lj}	$S'_f = \sigma_{aj}$	N_{Cj}

the reliability index of the component due to an equivalent random variable n_{i-j}. This random variable n_{i-j} has the same type of distribution and the same standard deviation as the cyclic loading n_{Lj}, but its mean is the equivalent cyclic number n_{eqi-j}.

Assumption Two: The equivalent cyclic number n_{eqi-j} is a deterministic cyclic number and will only affect the mean value of the distributed cyclic number at the cyclic stress level σ_{aj}. So the new distributed cyclic number including the equivalent cyclic number n_{eqi-j} at the cyclic stress level σ_{aj} will be $n_{eqi-j} + n_{Lj}$, which can be defined as n_{j-eq}. When compared with the original random variable n_{Lj}, the new random variable n_{j-eq} will have the same type of distribution and the same standard deviation but have an increase in its mean value by n_{eqi-j}.
     We will discuss two cases in the following sections: (1) both the component fatigue life N_C and the cyclic number n_L of the cyclic loading are the normal distribution; and (2) both the component fatigue life N_C and the cyclic number n_L of the cyclic loading are log-normal distribution.

Both Normal Distributions
When both the component fatigue life N_C and the cyclic number n_L of model #5 cyclic loading at the cyclic stress levels are normal distributions, as shown in Table 2.23. We can use the following procedure to determine the equivalent cyclic number n_{eqi-j} and the new distributed cyclic number n_{j-eq} at the stress level σ_{aj}. Then, we can calculate the reliability of a component under such cyclic loading spectrum.
     The limit state function of the component due to the cyclic loading at the stress level #i is:

$$g(N_{Ci}, n_{Li}) = N_{Ci} - n_{Li} = \begin{cases} > 0 & \text{Safe} \\ 0 & \text{Limit state} \\ < 0 & \text{Failure.} \end{cases} \qquad (2.50)$$

Since both the component fatigue life N_{Ci} (strength) and the number of cycles n_{Li} of the cyclic loading are normal distributions, the limit state function will be a normal distribution. The

Table 2.23: Normal distributions for n_L and N_C

Stress Level #	Model #5 Cyclic Loading Spectrum			Component Fatigue Life N_C		
	Stress Level	n_L (normal distribution)		Fatigue Strength	Distributed Fatigue Life N_C	
		Mean	Standard Deviation		Mean	Standard Deviation
i	σ_{ai}	$\mu_{n_{Li}}$	$\sigma_{n_{Li}}$	$S'_f = \sigma_{ai}$	$\mu_{N_{Ci}}$	$\sigma_{N_{Ci}}$
j	σ_{aj}	$\mu_{n_{Lj}}$	$\sigma_{n_{Lj}}$	$S'_f = \sigma_{aj}$	$\mu_{N_{Cj}}$	$\sigma_{N_{Cj}}$

reliability index of the limit state function (2.50) can be directly calculated by the following equation:

$$\beta_i = \frac{\mu_{N_{Ci}} - \mu_{Li}}{\sqrt{(\sigma_{N_{Ci}})^2 + (\sigma_{Li})^2}}. \tag{2.51}$$

Per Assumptions One and Two, the reliability index for the equivalent cyclic number n_{eqi-j} of the cyclic loading from the stress level #i to the stress level #j should have the same reliability index β_i:

$$\beta_i = \frac{\mu_{N_{Cj}} - n_{eqi-j}}{\sqrt{(\sigma_{N_{Cj}})^2 + (\sigma_{Lj})^2}}. \tag{2.52}$$

Rearrange Equation (2.52), the equivalent cyclic number n_{eqi-j} from the stress level #i to the stress level #j is:

$$n_{eqi-j} = \mu_{N_{Cj}} - \beta_i \sqrt{(\sigma_{N_{Cj}})^2 + (\sigma_{n_{Lj}})^2}. \tag{2.53}$$

Per Assumption Two, the new distributed cyclic number n_{j-eq} at the stress level σ_{aj} will have following mean and the same standard deviation:

$$\mu_{j-eq} = n_{eqi-j} + \mu_{Lj}$$
$$\sigma_{n_{j-eq}} = \sigma_{Lj}, \tag{2.54}$$

where μ_{j-eq} and $\sigma_{n_{j-eq}}$ are the mean and the standard deviation of the normally distributed random variable n_{j-eq}.

If there are more than two stress levels in the model #5 cyclic loading spectrum, we can use above procedures to continuously convert cyclic loading from one level to next level per Equations (2.51)–(2.54) until the equivalent cyclic loading in the last stress level has included all transferred equivalent cycles. If the stress level #j is the last stress level, the reliability of the component under model #5 cyclic loading spectrum is:

$$R = \Phi \left(\frac{\mu_{N_{Cj}} - n_{j-eq}}{\sqrt{(\sigma_{N_{Cj}})^2 + (\sigma_{n_{Lj}})^2}} \right), \tag{2.55}$$

where $\Phi\left(\cdot\right)$ is the CDF of standard normal distribution.

Example 2.15

A component is subjected to model #5 cyclic loading spectrum with three stress levels as listed in Table 2.24. Both the number of cycles n_L and the component fatigue life N_C follow normal distributions. Their distribution parameters at three stress levels are listed in Table 2.24. Calculate the reliability of this component.

Table 2.24: Normal distributions for n_L and N_C for Example 2.15

Stress Level #	Cyclic Stress Amplitude	Number of cycles n_L (normal distribution)		Component Fatigue Life N_C (normal distribution)	
		Mean ($\mu_{n_{Li}}$)	Standard Deviation ($\sigma_{n_{Li}}$)	Mean ($\mu_{N_{Ci}}$)	Standard Deviation ($\sigma_{N_{Ci}}$)
1	45 (ksi)	11,000	1,200	45,000	3,600
2	40 (ksi)	32,000	5,400	118,800	11,000
3	35 (ksi)	112,000	9,800	356,200	26,000

Solution:

The equivalent cyclic number n_{eq1-2} of the cyclic stress from the stress level 1 to the stress level 2 per Equations (2.51) and (2.53) is:

$$\beta_1 = \frac{\mu_{NC1} - \mu_{n_{L1}}}{\sqrt{\left(\sigma_{NC1}\right)^2 + \left(\sigma_{n_{L1}}\right)^2}} = \frac{45000 - 11000}{\sqrt{(3600)^2 + (1200)^2}} = 8.959787 \qquad (a)$$

$$n_{eq1-2} = \mu_{NC2} - \beta_1 \sqrt{\left(\sigma_{NC2}\right)^2 + \left(\sigma_{n_{L2}}\right)^2}$$

$$= 118800 - 8.959787 \times \sqrt{(11000)^2 + (5400)^2} = 9000.96. \qquad (b)$$

Therefore, the number of cycles n_{2-eq} in the stress level #2 including the transferred number of cycles from the stress level #1 will have the mean and standard deviation per Equation (2.54):

$$\mu_{2-eq} = n_{eq1-2} + \mu_{L_2} = 9000.96 + 32000 = 51006.96$$
$$\sigma_{2-eq} = \sigma_{L_2} = 5400. \qquad (c)$$

Now, repeat above calculations to convert the cyclic stress at the stress level #2 with the distribution parameters in Equation (c) to an equivalent number of cycles n_{eq2-3} at the stress level #3.

Per Equation (2.51), we have:

$$\beta_2 = \frac{\mu_{NC2} - \mu_{2-eq}}{\sqrt{(\sigma_{NC2})^2 + (\sigma_{2-eq})^2}} = \frac{118800 - 51006.96}{\sqrt{(11000)^2 + (5400)^2}} = 5.532329. \tag{d}$$

Per Equation (2.53)

$$n_{eq2-3} = \mu_{NC3} - \beta_2 \sqrt{(\sigma_{NC3})^2 + (\sigma_{L3})^2}$$

$$= 356200 - 5.532329 \times \sqrt{(26000)^2 + (9800)^2} = 202480.9. \tag{e}$$

Per Equation (2.54), the cyclic stress n_{3-eq} in the stress level #3, including all of the transferred number of cycles from the stress level #1 and #2 will have a following mean and standard deviation:

$$\mu_{3-eq} = n_{eq2-3} + \mu_{L3} = 202480.9 + 112000 = 314480.9$$

$$\sigma_{3-eq} = \sigma_{L3} = 9800. \tag{f}$$

Since the stress level #3 is the last stress level, we can use Equation (2.55) to calculate the reliability of the component by using the equivalent number of cycles n_{3-eq}, which includes all three cyclic stresses.

$$R = \Phi\left(\frac{\mu_{NC3} - \mu_{3-eq}}{\sqrt{(\sigma_{NC3})^2 + (\sigma_{L3})^2}}\right)$$

$$= \Phi\left(\frac{356200 - 314480.9}{\sqrt{(26000)^2 + (9800)^2}}\right) = \Phi(1.501465) = 0.9334. \tag{g}$$

∎

Both Lognormal Distributions

When both the component fatigue life N_C and the cyclic number n_L of model #5 cyclic loading at the cyclic stress levels are lognormal distributions, as shown in Table 2.25. We can use the following procedure to determine the equivalent cyclic number n_{eqi-j} and the new distributed cyclic number n_{j-eq} at the stress level σ_{aj}. Then, we can calculate the reliability of a component under model #5 cyclic loading spectrum.

For both lognormal distribution, the event $(N_{Ci} > n_{Li})$ is the same as the event $(\ln(N_{Ci}) > \ln(n_{Li}))$ because both N_{Ci} and n_{Li} are positive. The limit state function of the component due to

Table 2.25: Log-normal distributions for n_L and N_C

Stress Level #	Cyclic Stress Amplitude (constant)	Number of Cycles n_L (lognormal distribution)		Component Fatigue Life N_C (lognormal distribution)	
		Mean	Standard Deviation	Mean	Standard Deviation
i	σ_{ai}	$\mu_{\ln n_{Li}}$	$\sigma_{\ln n_{Li}}$	$\mu_{\ln N_{Ci}}$	$\sigma_{\ln N_{Ci}}$
j	σ_{aj}	$\mu_{\ln n_{Lj}}$	$\sigma_{\ln n_{Lj}}$	$\mu_{\ln N_{Cj}}$	$\sigma_{\ln N_{Cj}}$

the cyclic loading at the stress level #i is:

$$g\left(N_{Ci}, n_{Li}\right) = \ln\left(N_{Ci}\right) - \ln\left(n_{Li}\right) = \begin{cases} > 0 & \text{Safe} \\ 0 & \text{Limit state} \\ < 0 & \text{Failure.} \end{cases} \quad (2.56)$$

Now, both $\ln(N_{Ci})$ and $\ln(n_{Li})$ are a normal distribution. We can repeat the same calculations as those in Section 2.8.7. The reliability index of the limit state function (2.56) in the stress level #i is:

$$\beta_i = \frac{\mu_{\ln N_{Ci}} - \mu_{\ln Li}}{\sqrt{\left(\sigma_{\ln N_{Ci}}\right)^2 + \left(\sigma_{\ln Li}\right)^2}}. \quad (2.57)$$

The reliability index for the equivalent cyclic number n_{eqi-j} of the cyclic loading from the stress level #i to the stress level #j should have the same reliability index β_i.

$$\beta_i = \frac{\mu_{\ln N_{Cj}} - \ln\left(n_{eqi-j}\right)}{\sqrt{\left(\sigma_{N_{Cj}}\right)^2 + \left(\sigma_{Lj}\right)^2}}. \quad (2.58)$$

From Equation (2.58), the equivalent cyclic number n_{eqi-j} from the stress level i to the stress level j is:

$$n_{eqi-j} = \exp\left[\mu_{\ln N_{Cj}} - \beta_i \sqrt{\left(\sigma_{\ln N_{Cj}}\right)^2 + \left(\sigma_{\ln n_{Lj}}\right)^2}\right]. \quad (2.59)$$

Per Assumption Two in Section 2.8.7, n_{eqi-j} will be added to the mean of n_{Lj}. The mean of the lognormally distributed n_{Lj} with the log mean $\mu_{\ln n_{Lj}}$ and log standard deviation $\sigma_{\ln n_{Lj}}$ can be calculated by the following Equation (2.60) [8]:

$$\mu_{n_{Lj}} = \exp\left[\mu_{\ln n_{Lj}} + \frac{\left(\sigma_{\ln n_{Lj}}\right)^2}{2}\right]. \quad (2.60)$$

Therefore, the new mean for the new distributed cyclic number n_{j-eq} at the stress level σ_{aj} will be:

$$\mu_{j-eq} = n_{eqi-j} + \mu_{n_{Lj}} = n_{eqi-j} + \exp\left[\mu_{\ln n_{Lj}} + \frac{\left(\sigma_{\ln n_{Lj}}\right)^2}{2}\right]. \tag{2.61}$$

The new distributed cyclic number n_{j-eq} at the stress level σ_{aj} will still be a lognormal distribution. The log standard deviation of the new distributed cyclic number n_{j-eq} is the same as that of n_{Lj}.

$$\sigma_{\ln n_{j-eq}} = \sigma_{\ln L_j}. \tag{2.62}$$

For a lognormal distribution, the log mean of the new distributed cyclic number n_{j-eq} is:

$$\mu_{j-eq} = \exp\left[\mu_{\ln n_{j-eq}} + \frac{\left(\sigma_{\ln n_{j-eq}}\right)^2}{2}\right]. \tag{2.63}$$

Rearranging Equation (2.63), the log mean of the new distributed cyclic number n_{j-eq} is

$$\mu_{\ln n_{j-eq}} = \ln\left(\mu_{j-eq}\right) - \frac{\left(\sigma_{\ln n_{j-eq}}\right)^2}{2}, \tag{2.64}$$

where $\mu_{\ln n_{j-eq}}$ and $\sigma_{\ln n_{j-eq}}$ are the log mean and the log standard deviation of the lognormally distributed random variable n_{j-eq}.

If there are more than two stress levels in the model #5 cyclic loading spectrum, we can use above procedures to continuously convert cyclic stress from one level to next level per Equations (2.57)–(2.64) until the equivalent cyclic stress in the last stress level has included all transferred equivalent cycles. If the stress level j is the last stress level, the reliability of the component under model #5 cyclic loading spectrum is:

$$R = \Phi\left(\frac{\mu_{\ln N_{Cj}} - \mu_{\ln n_{j-eq}}}{\sqrt{\left(\sigma_{\ln N_{Cj}}\right)^2 + \left(\sigma_{\ln n_{j-eq}}\right)^2}}\right), \tag{2.65}$$

where $\Phi\left(\cdot\right)$ is the CDF of standard normal distribution.

Example 2.16
The component made of steel is subjected to model #5 cyclic loading spectrum with three fully reversed bending stress levels. Both the numbers of cycles of the fully reversed bending stress n_L and the component fatigue life N_C at each stress level are lognormal distributions as listed in Table 2.26. Calculate the reliability of the component under such cyclic loading spectrum.

Table 2.26: Lognormal distributions for n_L and N_C for Example 2.16

Stress Level #	Stress Amplitude (ksi)	Number of Cycles n_L (lognormal distribution)		Component Fatigue Life N_C (lognormal distribution)	
		$\mu_{\ln n_{Li}}$	$\sigma_{\ln n_{Li}}$	$\mu_{\ln N_{Ci}}$	$\sigma_{\ln N_{Ci}}$
1	66	10.8	0.19	12.86454	0.24868
2	86	8.9	0.18	10.85669	0.15658
3	106	8.1	0.16	9.44520	0.16809

Solution:

The equivalent cyclic number n_{eq1-2} of the cyclic stress from the stress level #1 to the stress level #2 per Equations (2.57) and (2.59) is:

$$\beta_1 = \frac{\mu_{\ln N_{C1}} - \mu_{\ln L1}}{\sqrt{\left(\sigma_{\ln N_{C1}}\right)^2 + \left(\sigma_{\ln L1}\right)^2}} = \frac{12.86454 - 10.8}{\sqrt{(0.24868)^2 + (0.19)^2}} = 6.596892 \qquad (a)$$

$$n_{eq1-2} = \exp\left[\mu_{\ln N_{C2}} - \beta_1 \sqrt{\left(\sigma_{\ln N_{C2}}\right)^2 + \left(\sigma_{\ln n_{L2}}\right)^2}\right]$$

$$= \exp\left[10.85699 - 6.596892 \times \sqrt{(0.15658)^2 + (0.18)^2}\right] = 10751.99, \qquad (b)$$

the new mean for the new distributed cyclic number n_{2-eq} at the stress level #2 per Equation (2.61) is:

$$\mu_{2-eq} = n_{eq1-2} + \exp\left[\mu_{\ln n_{L2}} + \frac{\left(\sigma_{\ln n_{L2}}\right)^2}{2}\right]$$

$$= 10751.99 + \exp\left[8.9 + \frac{(0.18)^2}{2}\right] = 18203.71. \qquad (c)$$

The log standard deviation of the new distributed cyclic number n_{2-eq} per Equation (2.62) is

$$\sigma_{\ln n_{2-eq}} = \sigma_{\ln L2} = 0.18 \qquad (d)$$

the log mean of the new distributed cyclic number n_{2-eq} per Equation (2.64) is

$$\mu_{\ln n_{2-eq}} = \ln\left(\mu_{2-eq}\right) - \frac{\left(\sigma_{\ln n_{2-eq}}\right)^2}{2} = \ln(18203,71) - \frac{(0.18)^2}{2} = 9.793181. \qquad (e)$$

Now, repeat above calculations to convert the cyclic stress at the stress level #2 with the distribution parameters in Equations (d) and (e) to an equivalent number of cycles n_{eq2-3} at the stress level #3.

Per Equation (2.57), the reliability index β_2 of the component under the new distributed cyclic number n_{2-eq} in the stress level #2 is:

$$\beta_2 = \frac{\mu_{\ln N_{C2}} - \mu_{\ln n_{2-eq}}}{\sqrt{\left(\sigma_{\ln N_{C2}}\right)^2 + \left(\sigma_{\ln L2}\right)^2}} = \frac{10.85699 - 9.793181}{\sqrt{(0.15658)^2 + (0.18)^2}} = 4.457785. \tag{f}$$

Per Equation (2.59), the equivalent cyclic number n_{eq2-3} from the stress level #2 to the stress level 3:

$$n_{eq2-3} = \exp\left[\mu_{\ln N_{C3}} - \beta_2 \sqrt{\left(\sigma_{\ln N_{C3}}\right)^2 + \left(\sigma_{\ln n_{L3}}\right)^2}\right]$$

$$= \exp\left[9.4452 - 4.457785 \times \sqrt{(0.16809)^2 + (0.16)^2}\right] = 4494.921; \tag{g}$$

the new mean for the new distributed cyclic number n_{3-eq} at the stress level #3 per Equation (2.61) is:

$$\mu_{3-eq} = n_{eq2-3} + \exp\left[\mu_{\ln n_{L3}} + \frac{\left(\sigma_{\ln n_{L3}}\right)^2}{2}\right]$$

$$= 4494.921 + \exp\left[8.1 + \frac{(0.16)^2}{2}\right] = 7831.83. \tag{h}$$

The log standard deviation of the new distributed cyclic number n_{3-eq} per Equation (2.62) is

$$\sigma_{\ln n_{3-eq}} = \sigma_{\ln L3} = 0.16; \tag{i}$$

the log-mean of the new distributed cyclic number n_{3-eq} per Equation (2.64) is

$$\mu_{\ln n_{3-eq}} = \ln\left(\mu_{3-eq}\right) - \frac{\left(\sigma_{\ln n_{3-eq}}\right)^2}{2} = \ln(7831.83) - \frac{(0.16)^2}{2} = 8.953151. \tag{j}$$

Since the stress level #3 is the last stress level, we can use Equation (2.65) to calculate the reliability of the component by using the equivalent number of cycles n_{3-eq}, which includes all three cyclic stresses:

$$R = \Phi\left(\frac{\mu_{\ln N_{C3}} - \mu_{\ln n_{3-eq}}}{\sqrt{\left(\sigma_{\ln N_{C3}}\right)^2 + \left(\sigma_{\ln n_{3-eq}}\right)^2}}\right) = \Phi\left(\frac{9.4452 - 8.953151}{\sqrt{(0.16809)^2 + (0.16)^2}}\right)$$

$$= \Phi(2.120303) = 0.9830.$$

■

2.8.8 RELIABILITY OF A COMPONENT UNDER MODEL #6 CYCLIC LOADING SPECTRUM

Model #6 cyclic loading spectrum is several distributed cyclic stress amplitudes at specified cycle numbers, that is, $(n_{Li}, \sigma_{ai}, i = 1, 2, \ldots)$. Here, n_{Li} is a constant number of cycles in the cyclic number level #i. Fully reversed cyclic stress level σ_{ai} in the cyclic number level #i is a distributed random variable. The corresponding component fatigue strength data will be the component fatigue strength S_{cfi} at the given fatigue life $N = n_{Li}$. The component fatigue strength S_{cfi} at the given fatigue life $N = n_{Li}$ can be calculated per Equation (2.29). Its mean and standard deviation can be calculated through Equations (2.30) and (2.31). This section will discuss how to calculate the reliability of a component under model #6 cyclic loading spectrum.

It is difficult to establish the limit state function of a component under model #6 cyclic loading spectrum. It is typically assumed that the influence of the sequence of cyclic loading on fatigue life or fatigue damage can be negligible. Therefore, each loading condition in the model #6 can be treated as an independent event. The author proposed an approach [17] in 2017 to estimate the reliability of a component under such cyclic loading spectrum. This approach has the following two assumptions.

Assumption One: Based on the concepts of the widely accepted Miner rule [7, 10, 18], the effect of the sequence of cyclic loading on the fatigue damage during the service life of the component can be ignored, so that each cyclic loading stress condition (n_{Li}, σ_{ai}) can be treated as an independent random event.

Assumption Two: Since the fatigue damage of the component due to these independent cyclic stress conditions is assumed to be independent, the estimation of the reliability R of the component under Model #6 (n_{Li}, σ_{ai}) is equal to the multiplication of the reliability R_i of the component under each cyclic loading condition (n_{Li}, σ_{ai}).

Assumption One is mainly based on the widely accepted linear cumulative fatigue damage theory. Assumption Two is a natural extension of Assumption One, but it is the expression of the reliability computational method. So, according to Assumption Two, the reliability of a component under the model #6 cyclic loading spectrum can be modeled as a series of reliability block diagrams, each of which represents the component under each cyclic loading condition. Thus, the reliability R of a component under model #6 cyclic loading spectrum is:

$$R = \prod_{i=1}^{L} R_i, \tag{2.66}$$

R_i is the reliability of the component under cyclic stress (n_{Li}, σ_{ai}) and can be calculated based on the following limit state function of the component under cyclic stress (n_{Li}, σ_{ai}):

$$g\left(S_{cfi}, \sigma_{ai}\right) = S_{cfi} - \sigma_{ai} = \begin{cases} > 0 & \text{Safe} \\ 0 & \text{Limit state} \\ < 0 & \text{Failure,} \end{cases} \qquad (2.67)$$

S_{cfi} and σ_{ai} are the component fatigue strength and the cyclic stress amplitude at the cyclic number level #i.

When both the fatigue strength S_{cfi} and the cyclic loading stress level σ_{ai} at the given cycle number n_{Li} are normal distributions, the reliability R_i can be directly calculated as the following:

$$R_i = P\left(S_{cfi} > \sigma_{ai}\right) = \Phi\left[\frac{\mu_{S_{cfi}} - \mu_{\sigma_{ai}}}{\sqrt{\left(\sigma_{S_{cfi}}\right)^2 + \left(\sigma_{\sigma_{ai}}\right)^2}}\right], \qquad (2.68)$$

where $\mu_{S_{cfi}}$ and $\sigma_{S_{cfi}}$ are the mean and the standard deviation of normally distributed S_{cfi}. $\mu_{\sigma_{ai}}$ and $\sigma_{\sigma_{ai}}$ are the mean and the standard deviation of normally distributed σ_{ai}. $\Phi\left(\cdot\right)$ is the CDF of standard normal distribution.

When both the fatigue strength S_{cfi} and the cyclic loading stress level σ_{ai} at the given cycle number n_{Li} are lognormal distributions, the reliability R_i can be directly calculated as the following:

$$R_i = P\left(S_{cfi} > \sigma_{ai}\right) = P\left[\ln(S_{cfi}) > \ln(\sigma_{ai}\right] = \Phi\left[\frac{\mu_{\ln S_{cfi}} - \mu_{\ln \sigma_{ai}}}{\sqrt{\left(\sigma_{\ln S_{cfi}}\right)^2 + \left(\sigma_{\ln \sigma_{ai}}\right)^2}}\right], \qquad (2.69)$$

where $\mu_{\ln S_{cfi}}$ and $\sigma_{\ln S_{cfi}}$ are the log-mean and the log standard deviation of log-normally distributed S_{cfi}. $\mu_{\ln \sigma_{ai}}$ and $\sigma_{\ln \sigma_{ai}}$ are the log-mean and the log standard deviation of log-normally distributed σ_{ai}. $\Phi\left(\cdot\right)$ is the CDF of standard normal distribution.

For a general case with any other type of distributions for S_{cfi} and σ_{ai}, the H-L, R-F method, or Monte Carlo method discussed in Chapter 3 can be used to calculate the reliability of the component per the limit state function Equation (2.67).

Example 2.17
A component is subjected to two distinguished fully reversed cyclic bending stresses due to two designed functions. The fully reversed cyclic bending stresses can be described by two normal distributed stress amplitudes at 8,000 cycles and 200,000 cycles, as shown in Table 2.27. The corresponding fatigue strength of the component at the given fatigue life 8,000 cycles and 200,000 cycles can be described as a normal distribution, as shown in Table 2.27. Calculate the reliability of the component.

Table 2.27: Distribution parameters of S_{cfi} and σ_{ai} for Example 2.17

Level #	Number of Cycles n_{Li} (constant)	Cyclic Stress Amplitude σ_{ai} (ksi) (normal distribution)		Component Fatigue Strength S_{cfi} (ksi) (normal distribution)	
		$\mu_{\sigma_{ai}}$	$\sigma_{\sigma_{ai}}$	$\mu_{S_{cfi}}$	$\sigma_{S_{cfi}}$
1	8,000	34.25	4.15	50.19	4.72
2	200,000	29.13	2.78	37.72	3.16

Solution:

Since both the component fatigue strength S_{cfi} and the cyclic stress amplitude σ_{ai} are normal distributions, the reliability R_1 of the component under cyclic stress σ_{a1} in the cyclic number level #1 per Equation (2.68) is:

$$R_1 = P\left(S_{cf1} > \sigma_{a1}\right) = \Phi\left[\frac{\mu_{S_{cf1}} - \mu_{\sigma_{a1}}}{\sqrt{\left(\sigma_{S_{cf1}}\right)^2 + \left(\sigma_{\sigma_{a1}}\right)^2}}\right]$$

$$= \Phi\left[\frac{50.19 - 34.25}{\sqrt{(4.72)^2 + (4.15)^2}}\right] = \Phi(2.536208) = 0.994397. \qquad (a)$$

Repeat the same calculation in the cyclic number level #2 per Equation (2.68),

$$R_2 = P\left(S_{cf2} > \sigma_{a2}\right) = \Phi\left[\frac{\mu_{S_{cf2}} - \mu_{\sigma_{a2}}}{\sqrt{\left(\sigma_{S_{cf2}}\right)^2 + \left(\sigma_{\sigma_{a2}}\right)^2}}\right]$$

$$= \Phi\left[\frac{37.72 - 29.13}{\sqrt{(3.16)^2 + (2.78)^2}}\right] = \Phi(2.040962) = 0.979373. \qquad (b)$$

Per Equation (2.66), the reliability of the component under this model #6 cyclic loading spectrum is:

$$R = \prod_{i=1}^{2} R_i = R_1 \times R_2 = 0.994397 \times 0.979373 = 0.9739. \qquad (c)$$

■

Example 2.18

A component is subjected to two distinguished fully reversed cyclic bending stresses due to two designed functions. The fully reversed cyclic bending stresses can be described by two normally distributed stress amplitudes at 5,000 cycles and 300,000 cycles, as shown in Table 2.28. The corresponding fatigue strength of the component at the given fatigue life 5,000 cycles and 300,000 cycles can be described as log-normal distribution, as shown in Table 2.28. Calculate the reliability of the component.

Table 2.28: Distribution parameters of S_{cfi} and σ_{ai} for Example 2.18

Level #	Number of Cycles n_{Li} (constant)	Cyclic Stress Amplitude σ_{ai} (ksi) (normal distribution)		Component Fatigue Strength S_{cfi} (ksi) (lognormal distribution)	
		$\mu_{\sigma_{ai}}$	$\sigma_{\sigma_{ai}}$	$\mu_{\ln S_{cfi}}$	$\sigma_{\ln S_{cfi}}$
1	5,000	54.2	6.775	4.3562	0.0321
2	300,000	45.2	5.3336	4.0507	0.0315

Solution:

Since the component fatigue strength S_{cfi} are lognormal distributions and the cyclic stress amplitude σ_{ai} are normal distributions, we need to use Equation (2.68) to establish the limit state functions.

The limit state functions for the cyclic number level #1 and #2 are

$$g\left(S_{cf1}, \sigma_{a1}\right) = S_{cf1} - \sigma_{a1} = \begin{cases} > 0 & \text{Safe} \\ 0 & \text{Limit state} \\ < 0 & \text{Failure} \end{cases} \tag{a}$$

$$g\left(S_{cf2}, \sigma_{a3}\right) = S_{cf2} - \sigma_{a2} = \begin{cases} > 0 & \text{Safe} \\ 0 & \text{Limit state} \\ < 0 & \text{Failure.} \end{cases} \tag{b}$$

Based on the limit state functions (a) and (b), the R-F method can be used to calculate their reliabilities. We can follow the procedure and the flowchart of the R-F method presented in Appendix A.2 to compile a MATLAB program. The iterative results for the limit state function (a) are listed in Table 2.29. From the iterative results, the reliability index β_1 and corresponding reliability R_1 of the component in this example are:

$$\beta_1 = 3.296752; \quad R_1 = \Phi(3.296752) = 0.999511.$$

Table 2.29: The iterative results at the cyclic number level #1

Iterative #	S_{cf1}^*	σ_{a1}^*	β^*	$\Delta|\beta^*|$
1	78.0005	78.0005	3.289598	
2	75.1051	75.1051	3.296735	0.007137
3	75.24278	75.24278	3.296752	1.7E-05

The iterative results for the limit state function (b) is listed in Table 2.30. From the iterative results, the reliability index β_2 and corresponding reliability R_2 of the component in this example are:

$$\beta_2 = 2.175236; \qquad R_2 = \Phi(2.175236) = 0.985194.$$

The reliability of the component under this cyclic loading spectrum per Equation (2.66) is:

$$R = \prod_{i=1}^{2} R_i = R_1 \times R_2 = 0.999511 \times 0.985194 = 0.9847.$$

■

Table 2.30: The iterative results at the cyclic number level #2

Iterative #	S_{cf2}^*	σ_{a2}^*	β^*	$\Delta|\beta^*|$
1	57.46615	57.46615	2.172718	
2	56.17362	56.17362	2.175234	0.002516
3	56.21165	56.21165	2.175236	2.25E-06

2.8.9   THE RELIABILITY OF A COMPONENT WITH P-S-N CURVES BY THE MONTE CARLO METHOD

The limit state function must be established first before the Monte Carlo method can be used to calculate the reliability of a component under cyclic loading Spectrum. When the P-S-N curves are used as the fatigue strength data, the limit state function of a component under six possible cyclic loading spectrums are listed here.

For model #1 cyclic loading spectrum (σ_a, n_L), which is a constant number of cycle n_L at a constant fully reversed cyclic stress amplitude σ_a, we can have two versions of limit state function per Equation (2.38) when the component fatigue strength S_{cf} is used or per Equation (2.40) when the component fatigue life N_C:

$$g\left(k_a, k_c,\ K_f, S_f'\right) = \frac{k_a k_b k_c}{K_f} S_f' - \sigma_a = \begin{cases} > 0 & \text{Safe} \\ 0 & \text{Limit state} \\ < 0 & \text{Failure} \end{cases} \qquad (2.38)$$

$$g\left(k_a, k_c, K_f, N\right) = N\left(\frac{k_a k_b k_c}{K_f}\right)^m - n_L = \begin{cases} > 0 & \text{Safe} \\ 0 & \text{Limit state} \\ < 0 & \text{Failure.} \end{cases} \qquad (2.40)$$

For model #2 cyclic loading spectrum (σ_a, n_L), which is a distributed number of cycle n_L at a constant fully reversed cyclic stress amplitude σ_a, the limit state function can be established per Equation (2.41) when the component fatigue life N_C at the given constant cyclic stress amplitude σ_a is provided. The limit state function will be constructed per Equation (2.42) when the material fatigue life N at the given constant cyclic stress amplitude σ_a is provided:

$$g\left(N_c, n_L\right) = N_c - n_L = \begin{cases} > 0 & \text{Safe} \\ 0 & \text{Limit state} \\ < 0 & \text{Failure} \end{cases} \qquad (2.41)$$

$$g\left(k_a, k_c, K_f, N, n_L\right) = N\left(\frac{k_a k_b k_c}{K_f}\right)^m - n_L = \begin{cases} > 0 & \text{Safe} \\ 0 & \text{Limit state} \\ < 0 & \text{Failure.} \end{cases} \qquad (2.42)$$

For model #3 cyclic loading spectrum (σ_a, n_L), which is a distributed fully reversed cyclic stress amplitude σ_a, with a constant number of cycles n_L, the limit state function can be established per Equation (2.43) when the component fatigue strength S_{Cf} at the given constant fatigue life $N = n_L$ is provided. The limit state function will be constructed per Equation (2.44) when the material fatigue strength S_f' at the given constant fatigue life $N = n_L$ is provided:

$$g\left(S_{Cf}, \sigma_a\right) = S_{Cf} - \sigma_a = \begin{cases} > 0 & \text{Safe} \\ 0 & \text{Limit state} \\ < 0 & \text{Failure} \end{cases} \qquad (2.43)$$

$$g\left(k_a, k_c, K_f, S_f', \sigma_a\right) = \frac{k_a k_b k_c}{K_f} S_f' - \sigma_a = \begin{cases} > 0 & \text{Safe} \\ 0 & \text{Limit state} \\ < 0 & \text{Failure.} \end{cases} \qquad (2.44)$$

For model #6 cyclic loading spectrum $(\sigma_{ai}, n_{Li}, \; i = 1, 2, \ldots)$, which are distributed fully re-versed cyclic stress amplitude σ_{ai} with a constant number of cycles n_{Li} in the cyclic number level #i, the limit state function of the component for each cyclic number level #i can be established per Equation (2.67):

$$g\left(S_{cfi}, \sigma_{ai}\right) = S_{cfi} - \sigma_{ai} = \begin{cases} > 0 & \text{Safe} \\ 0 & \text{Limit state} \quad i = 1, 2, \ldots \\ < 0 & \text{Failure.} \end{cases} \qquad (2.67)$$

For all the above cases, we can directly use the Monte Carlo method to calculate the reliability of a component per its limit state function. The Monte Carlo method procedure and program flowchart is displayed in Appendix A.3.

For model #4 and model #5 cyclic loading spectrums $(\sigma_{ai}, n_{Li}, \; i = 1, 2, \ldots)$, we could not build their limit state functions, but the equivalent fatigue damage concepts could be used to calculate the reliability.

The author presented an approach [19] in 2018 to use the Monte Carlo method to calculate the reliability of a component under these two cyclic loading spectrums. The two key concepts in the widely accepted Miner rule are that fatigue damage caused by cyclic loading could be treated as independent random events and could be cumulated linearly. Based on these two key concepts in the Miner rule, the following is the computational algorithm for implementing the Monte Carlo method to calculate the reliability of a component under the model #4 and #5 cyclic loading spectrum.

The accumulated fatigue damage F_j in the jth trial of the Monte Carlo method is:

$$F_j = \sum_{i=1}^{i=I} \frac{n_{Lij}}{N_{Cij}}, \qquad (2.70)$$

where n_{Lij} is a randomly generated sample of the distributed number of cycles n_{Li} at the cyclic stress level σ_{ai} in the jth trial for the model #5 cyclic loading spectrum. n_{Lij} will be equal to n_{Li} for the model #4 cyclic loading spectrum. N_{Cij} is a randomly generated sample of the distributed component fatigue life N_{Ci} at the cyclic stress level σ_{ai} in the jth trial. The subscript i represents the ith cyclic stress level. The symbol I represents the total number of different cyclic stress levels. The trial result tn_j of the jth Monte Carlo simulation will be determined per the following equation:

$$tn_j = \begin{cases} 1, & \text{if } F_j < 1 \\ 0, & \text{if } F_j \geq 1. \end{cases} \qquad (2.71)$$

In Equation (2.71), the trial result $tn_j = 1$ represents that the component in the jth trail is safe because the cumulative damage F_j is less than 1. The trial result $tn_j = 0$ indicates that the component in the jth trail fails because the cumulative damage F_j is larger than 1. The sum of all trial results $\sum tn_j$ will be the number of trails with a safe status. So, the component reliability R under such cyclic loading spectrum is:

$$R = \left(\sum_{j=1}^{N_t} tn_j \right) / N_t, \tag{2.72}$$

where N_t is the total number of trials in the Monte Carlo simulation. Since the limit state function of a component under cyclic loading spectrum is typically not very complicated, $N_t = 15{,}998{,}400$ can be used.

Example 2.19
Use the Monte Carlo method to calculate the reliability of Example 2.14 in Section 2.8.6. A component is subjected to cyclic loading at three different constant fully reversed cyclic stress levels with three different constant numbers of cycles as listed in Table 2.31. The component fatigue life N_c under the corresponding stress levels are also listed in Table 2.31. Use the Monte Carlo method to calculate the reliability of the component.

Table 2.31: Model #4 cyclic loading spectrum and corresponding component fatigue life for Example 2.19

Stress Level i	Cyclic Stress Amplitude σ_{ai} (psi)	Number of Cycles n_{Li}	Component Fatigue Life N_{Ci} at σ_{ai} (psi) (lognormal distribution)	
			$\mu_{\ln N_{Ci}}$	$\sigma_{\ln N_{Ci}}$
1	65,000	81,000	12.95987	0.198
2	85,000	16,000	11.01311	0.197
3	105,000	2,800	9.47966	0.195

Solution:
We can use the above Equations (2.70), (2.71), and (2.72) to compile the Monte Carlo method program, which is displayed in Appendix A.3. Based on the proposed computational algorithm, the MATLAB program can be compiled, and the reliability of the component from the MATLAB program is:

$$R = \left(\sum_{j=1}^{N_t} tn_j \right) / N_t = \frac{15992388}{15998400} = 0.9996.$$

The result from Example 2.14 by using the equivalent fatigue damage concept is $R = 0.9877$. The result of the Monte Carlo method is 0.9996. The results are not the same. However, the relative difference is only 2.1%.   ∎

Example 2.20
Use the Monte Carlo method to calculate the reliability of the component in Example 2.15 in Section 2.8.7. A component is subjected to model #5 cyclic loading stresses with three stress levels as listed in Table 2.32. The component fatigue life at a corresponding cyclic stress level is also listed in Table 2.32. Use the Monte Carlo method to calculate the reliability of this component.

Table 2.32: Model #5 cyclic loading spectrum and corresponding component fatigue life for Example 2.20

Stress Level #	Cyclic Stress Amplitude	Number of Cycles n_{Li} (normal distribution)		Component Fatigue Life N_{Ci} (normal distribution)	
		$\mu_{n_{Li}}$	$\sigma_{n_{Li}}$	$\mu_{N_{Ci}}$	$\sigma_{N_{Ci}}$
1	45 (ksi)	11,000	1,200	45,000	3,600
2	40 (ksi)	32,000	5,400	118,800	11,000
3	35 (ksi)	112,000	9,800	356,200	26,000

Solution:
We can use the above Equations (2.70), (2.71), and (2.72) to compile the Monte Carlo method program. Based on the proposed computational algorithm, the MATLAB program can be compiled, and the reliability of the component from the MATLAB program is:

$$R = \left(\sum_{j=1}^{N_t} tn_j \right) / N_t = \frac{15763743}{15998400} = 0.9853.$$

The result from Example 2.15 by using the equivalent fatigue damage concept is $R = 0.9334$. The result of the Monte Carlo method is 0.9853. The results are not the same. However, the relative difference is only 5.6%. ∎

2.9 THE PROBABILISTIC FATIGUE DAMAGE THEORY (THE K-D MODEL)

2.9.1 INTRODUCTION

For high-cycle fatigue issue, which is the main focus of this book, most of the fatigue test data are from fatigue tests under a constant cyclic stress level. There are many theories to interpret and to describe fatigue data [1, 20]. One of the typical well-known approaches for describing the fatigue test data is the P-S-N curve approach, which has been discussed in Section 2.8 [3–5].

In this approach, a probabilistic distribution function is used to describe the fatigue test data at the same cyclic stress level. If fatigue tests are at several cyclic stress levels such as seven stress levels, there will be seven different probabilistic distribution functions if there are big enough number of tests in each stress levels. However, the P-S-N curve approach has the following four issues in its implementation for fatigue reliability design.

1. Since fatigue tests are time-consuming, there are only a few fatigue test data at each stress levels, which are common cases, as shown in the fatigue data book [21]. In such a situation, the P-S-N curve cannot be constructed due to the small sample size.

2. In some available fatigue test data, the total number of fatigue test might be more than 30 even though the number of fatigue tests in the same stress level is small, which is the common case in the fatigue data book [21]. The P-S-N curve approach cannot use such data to construct the P-S-N curves.

3. When the cyclic stress level in cyclic loading is not equal to the fatigue test stress level, the probabilistic distribution function at this level is not available in the P-S-N curves, which is a typical case in reality for fatigue design. So the P-S-N curves cannot be used to solve this type of issue. The P-S-N curve approach could use the interpolation method to obtain the probabilistic distribution function at the required stress level for reliability fatigue design. However, this distribution function is not directly obtained from or based on the test data, and it might induce some big error.

4. In fatigue tests, actual dimensions of fatigue specimen will be slightly different. Therefore, the actual stress level for the same nominal stress level fatigue test might be different; even the nominal stress level is the same. However, the P-S-N curve approach ignores these differences and use the nominal stress level to create the P-S-N curves.

The fatigue damage mechanism, which has been discussed in Section 2.2, shows that the fatigue damage is mainly caused by cyclic loading and randomly distributed defects inside a component such as voids and dislocations and or on the surface of a component such as manufacturing scratches. This result strongly suggests that the fatigue damage mechanism for the same type of material specimen under different cyclic test stress levels should be the same. Therefore, we can use all test data from every stress level to construct a probabilistic fatigue damage model for presenting material strength, which is the topic in this Section 2.9.

2.9.2   THE MATERIAL FATIGUE STRENGTH INDEX K_0

In the traditional fatigue design, the S-N curve is typically plotted in a logarithmetic axis with a fatigue strength S_f' verse the fatigue life N. S_f' is equal to a fully reversed stress amplitude σ_a. The fatigue life N is the number of cycles to failure at the stress level σ_a. This traditional S-N curve in logarithmic axes will typically be treated as a straight lineper Equation (2.1):

$$N(S_f')^m = Constant. \tag{2.1}$$

In the traditional S-N curve, $N(S'_f)^m$ is treated as constants. It is obvious that the $N(S'_f)^m$ cannot be a constant when the reliability of a component is used as the design parameter. The author in 1993 proposed a probabilistic fatigue damage model [22–24] in which $N(S'_f)^m$ is treated as a random variable and is called as the material fatigue strength index.

The material fatigue strength index K_0 is a mechanical property of a material and is solely determined by experimental fatigue data per Equation (2.73) and can be used to indirectly represent the materials fatigue resistance to the fatigue damage or the material fatigue strength. The material fatigue strength index K_0 is a random variable. Its sample value can be calculated by fatigue test results of material fatigue specimen from the same type of cyclic loading stress, which could be cyclic bending stress, or cyclic axial stress, or cyclic torsion stress.

$$K_0 = N_{ij} (\sigma_{ai})^m,\tag{2.73}$$

where the subscript i represents the ith fatigue stress level σ_{ai}. The subscript j represents the jth fatigue test in the ith fatigue stress level σ_{ai}. (N_{ij}, σ_i) are the fatigue test results of the jth fatigue test at the ith fatigue stress level σ_{ai}. N_{ij} is the number of cycles to failure or the material fatigue life of the jth fatigue test at the ith fatigue stress level σ_{ai}. σ_{ai} is a fully reversed cyclic stress amplitude or the material fatigue strength, which is equal to the fully reversed cyclic stress amplitude of the ith fatigue stress level σ_{ai}. m is a material fatigue property, is the slope of the traditional S-N curve and can be calculated per Equation (2.2) which has been discussed in Section 2.3 and repeated here as Equation (2.74):

$$m = \frac{I \sum_{i}^{I} [\ln(\sigma_{ai}) \cdot \ln(N_i)] - \sum_{i}^{I} [\ln(\sigma_{ai})] \sum_{i}^{I} [\ln(N_i)]}{I \sum_{i}^{I} [\ln(\sigma_{ai})]^2 - \left[\sum_{i}^{I} \ln(\sigma_{ai})\right]^2},\tag{2.74}$$

where I is the number of different stress amplitude σ_{ai} for the total fatigue tests; σ_{ai} is the ith stress amplitude of a fully reversed cyclic stress in the fatigue test; $\ln(N_i)$ is the average fatigue life in log-scale at the fatigue test level σ_{ai}, which can be calculated per Equation (2.3) and repeated here as (2.75):

$$\ln(N_i) = \frac{\sum_J \ln(N_{ij})}{J},\tag{2.75}$$

where N_{ij} is the number of cycles to the failure of the jth fatigue test under the ith same stress amplitude σ_{ai}.

For the same type of cyclic loading test stress, there are three possible cyclic loading stresses: fully reversed cyclic stress, non-zero mean cyclic stress, and notched cyclic stress.

For a fatigue test under a fully reversed cyclic stress level σ_a, σ_{ai} will be just equal to the fully reversed cyclic stress amplitude σ_a:

$$\sigma_{ai} = \sigma_a \qquad \text{for a fully reverse cyclic stress.}\tag{2.76}$$

If a fatigue test is conducted under a non-zero mean cyclic stress level (σ_a, σ_m), it should be converted into fully reversed cyclic stress. The modified Goodman approach can be used to consider the effect of mean stress per Equation (2.21), which has been discussed in Section 2.5 and repeated here as Equation (2.77):

$$\sigma_{ai} = \begin{cases} \sigma_a & \text{when } \sigma_m < 0 \\ \sigma_a \dfrac{S_u}{(S_u - \sigma_m)} & \text{when } \sigma_m \geq 0, \end{cases} \tag{2.77}$$

where S_u is the material ultimate tensile strength and will be treated as a deterministic value because it is only used for considering the effect of mean stress.

For a notched fatigue test under the cyclic stress level (σ_a, σ_m) with a fatigue stress concentration factor K_f, the cyclic stress level needs to be transferred into a fully revered cyclic stress. It is typically that K_f will be multiplied with the stress amplitude. σ_{ai} can be calculated by the following equation:

$$\sigma_{ai} = \begin{cases} K_f \sigma_a & \text{when } \sigma_m < 0 \\ K_f \sigma_a \dfrac{S_u}{(S_u - \sigma_m)} & \text{when } \sigma_m \geq 0, \end{cases} \tag{2.78}$$

where K_f is the fatigue stress concentration factor, which has been discussed in Section 2.6. S_u is the material ultimate tensile strength.

For each fatigue test, we can obtain one sample value of the material fatigue strength index K_0 per Equation (2.73). When the number of fatigue test is big enough such as more than 30 tests, we can plot its histogram. Based on the shape of the histogram, we can assume its type of distribution function and then conduct the goodness-of-fit test to verify the assumption. Finally, we can determine its type of distribution and corresponding distribution parameters. These topics have been discussed in Section 2.13 of Le [8]. The material fatigue strength index K_0 is typically a lognormal distribution [22–24]. Since the sample value of K_0 is calculated per Equation (2.73) and is related to the value of m, m will be one parameter for describing the material fatigue strength index K_0. If we assume that the material fatigue strength index K_0 is a lognormal distribution with a log-mean $\mu_{\ln K_0}$ and a log standard deviation $\sigma_{\ln K_0}$, the material fatigue strength index K_0 will be a three-parameter distribution as shown in Table 2.33.

Table 2.33: The material fatigue strength index K_0 with three distribution parameters

Slope of the Traditional S-N Curve	Log-normally Distributed K_0	
	The Log Mean	The Log Standard Deviation
m	$\mu_{\ln K_0}$	$\sigma_{\ln K_0}$

Now, we will use the fatigue test data listed in Table 2.11 in Section 2.8.1 to show how to get the distribution parameters of the material fatigue strength index K_0.

Example 2.21

A sheet-type flat fatigue specimen designed per ASTM STM E466-15 is shown in Figure 2.8. The material is aluminum 6061-T6 10 Gauge sheet. Its chemical composition is shown in Table 2.9. Its mechanical properties are shown in Table 2.10. For all fatigue test specimen, the nominal dimensions of the middle section of the specimen are the width $b = 0.600 \pm 0.005''$ and the thickness $t = 0.100 \pm 0.005''$. The fatigue test loading is cyclic axial loading with a loading frequency 20 (Hz), and the loading ratio $S_r = \frac{F_{min}}{F_{max}} = 0$. The test conditions and results of these five different cyclic stress levels are listed in Table 2.11. Stresses in Table 2.11 are calculated by using the nominal dimensions, that is, the width $b = 0.600''$ and the thickness $t = 0.100''$. Use those data to determine the type of distribution of the material fatigue strength index K_0 and its three distribution parameters.

Solution:

(1) The slope of the traditional S-N curve.

To calculate the sampling value of each fatigue test, we need to use Equation (2.74) to calculate the slope of the traditional S-N curve. Since the cyclic stresses in these fatigue tests are no-zero-mean cyclic stresses, we need to use Equation (2.67) to convert them into equivalent fully reversed cyclic stresses. We also need to calculate the average fatigue life in log-scale at the fatigue test level σ_{ai} per Equation (2.75). The equivalent fully revered cyclic stress amplitude σ_{ai} and the average fatigue life in log-scale $\ln(N_i)$ at the ith fatigue test level σ_{ai} per the test data are listed in Table 2.34. In Table 2.34, the first column is the cyclic stress level #i. The 2nd and 3rd columns are the mean stress and stress amplitude of the cyclic axial stress at the corresponding cyclic stress level. The fourth column is the number of tests at the same cyclic stress level. The fifth column is the equivalent fully revered cyclic stress amplitude. The sixth column is the equivalent fully revered cyclic stress amplitude in a log-scale. The seventh column is the average fatigue life in log-scale at the same cyclic stress level.

Use the data in the 6th and 7th columns in Table 2.34 per Equation (2.74) to conduct the linear regression in Excel. m value is displayed in Figure 2.16:

$$m = 3.8812. \tag{a}$$

(2) The sampling data of the material fatigue strength index K_0.

In this example, we have five different stress levels. Since all cyclic stresses are the same type of cyclic stress, that is, the cyclic axial stress, we can use all of 195 fatigue tests in all 5 stress levels to calculate the sampling values of the material fatigue strength index K_0. Based on the test data (N_{ij}, σ_{ai}) listed in Table 2.11, we can get 195 sampling data of K_0 per Equation (2.73).

Table 2.34: The equivalent stress amplitude and the average of $\ln(N)$

Stress level # i	σ_a (ksi)	σ_m (ksi)	Number of tests (J)	Equivalent stress σ_{ai} (ksi)	$\ln(\sigma_{ai})$	$\ln(N_i)$
1	20.833	20.833	50	35.12623	3.558948	11.4736
2	22.083	22.083	55	38.83228	3.659252	11.0410
3	22.5	22.5	30	40.13937	3.692358	10.9551
4	22.917	22.917	30	41.48497	3.725331	10.7831
5	23.333	23.333	30	42.87081	3.758191	10.716

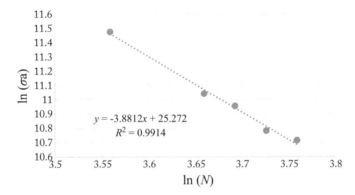

Figure 2.16: The traditional S-N curve.

(3) The histogram.

The histogram of K_0 with a total of 195 sampling data is displayed in Figure 2.17. From the histogram shown in Figure 2.17, the material fatigue strength index K_0 seems to be a log-normally distributed random variable.

(4) Type of distribution and its distribution parameters.

Following the procedure discussed in Section 2.13.3 of Le [8], we can create the MAT-LAB program to run the Chi-square (χ^2) goodness-of-fit test. The test result proves that the material fatigue strength index K_0 in this example can be described by a lognormal distribution. The three distribution parameters for the material fatigue strength index K_0 of this example are listed in Table 2.35. ∎

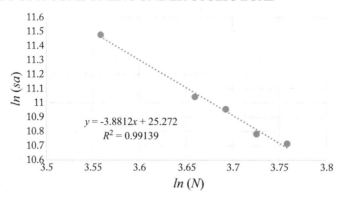

Figure 2.17: The histogram of K_0 with a sampling size 195.

Table 2.35: K_0 with three distribution parameters

Material: 6061-T6 10-Gauge Sheet	Sample Size: 195	Test Conditions: Cyclic Axial Stress, Milled Machined Sheet-type Flat Specimen	
The Slope of the Traditional S-N Curve: m	The Log-normally Distributed K_0		
	The Log Mean: $\mu_{\ln K_0}$	The Log Standard Deviation: $\sigma_{\ln K_0}$	
3.8812	25.3014	0.245451	

2.9.3   THE COMPONENT FATIGUE STRENGTH INDEX K

As we have discussed in Section 2.4, we will use the Marin modification factors to consider the differences between material fatigue specimen and component. After the differences are considered, we can obtain the component fatigue strength index K.

The component fatigue strength index K is a mechanical property of a component and is obtained through modifying the material fatigue strength index K_0 by the Martin modification factors per Equation (2.79), and can be used to indirectly represent the component fatigue resistance to the fatigue damage or the component fatigue strength. The component fatigue strength index K can be determined by the following equation:

$$K = N_i \left(k_a k_b k_c \sigma_{ai} \right)^m = \left(k_a k_b k_c \right)^m \left[N_i \left(\sigma_{ai} \right)^m \right] = \left(k_a k_b k_c \right)^m K_0, \qquad (2.79)$$

where k_a is the surface finish modification factor. k_b is the size modification factor. k_c is the loading modification factor. These Martin modification factors and its calculations have been discussed in Section 2.3. m is the slope of the traditional S-N curve and can be determined by Equation (2.74). K_0 is the material fatigue strength index and can be determined by Equation (2.73).

Equation (2.79) can be directly used in the limit state function for the reliability calculation. We can also typically assume that the component fatigue strength index K will have the same type of distribution, that is, a lognormal distribution, as that for material fatigue strength index K_0. We can use the following equations to calculate the log mean $\mu_{\ln K}$ and the log standard deviation $\sigma_{\ln K}$ of the component fatigue strength index K:

$$\mu_{\ln K} = m \times \ln\left(\mu_{k_a} k_b \mu_{k_c}\right) + \mu_{\ln K0} \tag{2.80}$$

$$\sigma_{\ln K} = \sqrt{m^2\left[\left(\frac{\sigma_{k_a}}{\mu_{k_a}}\right)^2 + \left(\frac{\sigma_{k_c}}{\mu_{k_c}}\right)^2\right] + (\sigma_{\ln K0})^2}, \tag{2.81}$$

where μ_{k_a} and σ_{k_a} are the mean and the standard deviation of the normally distributed k_a. μ_{k_c} and σ_{k_c} are the mean and the standard deviation of the normally distributed k_c. k_b is treated as a deterministic value. Their calculations have been discussed in Section 2.4.

The component fatigue strength index K has three distribution parameters. The slope of the traditional S-N curve for K will be the same as that for K_0. The component fatigue strength index K is still a log-normally distributed random variable. Its log mean and log standard deviation will be calculated per Equations (2.80) and (2.81). The three distribution parameters for the component fatigue strength index K are listed in Table 2.36.

Table 2.36: The component fatigue strength index K with three distribution parameters

Slope of the Traditional S-N Curve	Log-normally Distributed K	
	Log Mean	The Log Standard Deviation
m	$\mu_{\ln K}$	$\sigma_{\ln K}$

2.9.4   THE COMPONENT FATIGUE DAMAGE INDEX D

After the definition of the component fatigue strength index K and its calculation equations are explained, we can define the component fatigue damage index and its calculation equation accordingly.

The component fatigue damage index is an indirect measurement of the component fatigue damage under cyclic loading, is linearly accumulated fatigue damage by each cycle of cyclic stress and can be calculated by cyclic loading per Equation (2.82). Based on the definition and the calculation Equation (2.79) for the component fatigue strength index K, the fatigue damage index D of a component is calculated by the following Equation (2.82):

$$D = \sum_{i=1}^{L} n_{Li} \left(K_f \sigma_{ai}\right)^m, \tag{2.82}$$

where L is the number of different cyclic loading levels and m is a material fatigue property and is the slope of the traditional S-N curve. m is determined per Equation (2.74). K_f is the fatigue stress concentration factor on the component critical section, which has been discussed and can be calculated per Equations (2.22)–(2.25) in Section 2.6. n_{Li} is the number of cycles of the ith cyclic loading stress level σ_{ai}. σ_{ai} is an equivalent fully reversed cyclic stress amplitude and can be calculated by the following equation:

$$\sigma_{ai} = \begin{cases} \sigma_a & \text{for a fully reversed cyclic stress} \\ \sigma_a & \sigma_m \leq 0 \text{ of non-zero mean cyclic stress} \\ \sigma_a \dfrac{S_u}{(S_u - \sigma_m)} & \sigma_m > 0 \text{ of non-zero mean cyclic stress,} \end{cases} \tag{2.83}$$

where (σ_a, σ_m) are the stress amplitude and the mean stress of cyclic stress. S_u is the ultimate material strength. Equation (2.83) is based on the modified Goodman approach for the consideration of the effect of mean stress in cyclic stress.

n_{Li} or σ_{ai} can be a constant value or a distributed random variable, which is defined by the provided cyclic loading spectrum.

Following are the equations for calculating component fatigue damage per given cyclic spectrum.

For model #1, model #2, and model #3 cyclic loading spectrum (σ_a, n_L), the component fatigue damage due to the cyclic loading spectrums is:

$$D = n_L \left(K_f \sigma_a\right)^m, \tag{2.84}$$

where K_f and m have the same meaning as those in Equation (2.82). For model #1, n_L is a constant number of cycles and σ_a is the fully reversed constant cyclic stress amplitude. For the model #2, n_L is a distributed number of cycles and σ_a is the fully reversed constant cyclic stress amplitude. For model #3, n_L is a constant number of cycles and σ_a is the fully reversed distributed cyclic stress amplitude.

For model #4, model #5, and model #6 cyclic loading spectrum $(\sigma_{ai}, n_{Li}, i = 1, 2, \ldots, L)$, the component fatigue damage due to the cyclic loading spectrums is calculated per Equation (2.82) and repeated here as Equation (2.85):

$$D = \sum_{i=1}^{L} n_{Li} \left(K_f \sigma_{ai}\right)^m, \tag{2.85}$$

where K_f, m, and L have the same meaning as those in Equation (2.82). For model #4, n_{Li} is a constant number of cycles and σ_{ai} is the fully reversed constant cyclic stress amplitude at the ith cyclic stress level. For model #5, n_{Li} is a distributed number of cycles and σ_{ai} is the fully reversed constant cyclic stress amplitude at the ith cyclic stress level. For model #6, n_{Li} is a constant number of cycles and σ_{ai} is a distributed fully reversed constant cyclic stress amplitude at the ith cyclic stress level.

2.9.5 THE PROBABILISTIC FATIGUE DAMAGE THEORY (THE K-D MODEL)

Based on the definitions of the component fatigue strength index K and the component fatigue damage index D, the general limit state function of a component under cyclic loading is:

$$g(K, D) = K - D = \begin{cases} > 0 & \text{Safe} \\ 0 & \text{Limit state} \\ < 0 & \text{Failure.} \end{cases} \qquad (2.86)$$

If the material fatigue strength index K_0 is given, the limit state function of a component under model #1, model #2, or model #3 cyclic loading spectrum is:

$$g(K, D) = K - D = (k_a k_b k_c)^m K_0 - n_L \left(K_f \sigma_a\right)^m = \begin{cases} > 0 & \text{Safe} \\ 0 & \text{Limit state} \\ < 0 & \text{Failure.} \end{cases} \qquad (2.87)$$

All variables in Equation (2.87) have the same meanings as those in Equations (2.79) and (2.84).
   If the material fatigue strength index K_0 is given, the limit state function of a component under model #4, model #5, or model #6 cyclic loading spectrum is:

$$g(K, D) = K - D = (k_a k_b k_c)^m K_0 - \sum_{i=1}^{L} n_{Li} \left(K_f \sigma_{ai}\right)^m = \begin{cases} > 0 & \text{Safe} \\ 0 & \text{Limit state} \\ < 0 & \text{Failure.} \end{cases} \qquad (2.88)$$

All variables in Equation (2.88) have the same meanings as those in Equations (2.79) and (2.85).
   When this probabilistic fatigue damage model (the K-D model) is used, the reliability of a component under any cyclic loading spectrum can be calculated per Equations (2.87) or (2.88).

2.9.6 RELIABILITY OF A COMPONENT UNDER CYCLIC AXIAL LOADING

Per Equations (2.87) or (2.88), we can establish the limit state function of a component under any type of cyclic axial loading spectrum. After the limit state function of a component under cyclic axial loading is established, we can use the H-L method, R-F method, or Monte Carlo method to calculate its reliability. In this section, we will use two examples to demonstrate how to calculate the reliability of a component under cyclic axial loading spectrum.

Example 2.22
A constant round bar with a diameter $0.850 \pm 0.005''$ is subjected to model #1 cyclic axial loading spectrum as listed in Table 2.37. The ultimate material strength S_u is 75 (ksi). Three parameters

of the component fatigue strength index K are $m = 8.21$, $\mu_{\ln K} = 41.738$, and $\sigma_{\ln K} = 0.357$. For the component fatigue strength index K, the stress unit is ksi. Calculate the reliability of this bar.

Table 2.37: Model #2 cyclic axial loading for Example 2.22

Number of Cycles n_L	The Mean of the Cyclic Axial Loading F_m (klb)	The Amplitude of Cyclic Axial Loading F_a (klb) (normal distribution)	
		μ_{F_a}	σ_{F_a}
103000	8.85	14.11	1.51

Solution:

(1) The cyclic axial stress and the component fatigue damage index.

The mean stress σ_m and the stress amplitude σ_a of the bar due to the cyclic axial loading are:

$$\sigma_m = \frac{F_m}{\pi d^2/4} = \frac{4F_m}{\pi d^2} \tag{a}$$

$$\sigma_a = \frac{F_a}{\pi d^2/4} = \frac{4F_a}{\pi d^2}. \tag{b}$$

Since the cyclic axial stress is a non-zero mean cyclic stress, we need to convert it into a fully reversed cyclic axial stress per Equation (2.83). The equivalent stress amplitude of this converted fully reversed cyclic stress is:

$$\sigma_{a-eq} = \sigma_a \frac{S_u}{(S_u - \sigma_m)} = \frac{4F_a S_u}{(\pi S_u d^2 - 4F_m)}. \tag{c}$$

The component fatigue damage index D of the bar under model #2 cyclic loading spectrum per Equation (2.84) is

$$D = n_L \left(K_f \sigma_a\right)^{8.21} = n_L \left[\frac{4F_a S_u}{(\pi S_u d^2 - 4F_m)}\right]^{8.21}. \tag{d}$$

(2) The limit state function of this bar.

The limit state function of the bar per Equation (2.87) is

$$g(K, F_a, d) = K - n_L \left[\frac{4 F_a S_u}{(\pi S_u d^2 - 4 F_m)} \right]^{8.21}$$

$$= K - 103000 \left[\frac{300 F_a}{(75 \pi d^2 - 35.4)} \right]^{8.21} = \begin{cases} > 0 & \text{Safe} \\ 0 & \text{Limit state} \\ < 0 & \text{Failure.} \end{cases} \qquad (e)$$

In this limit state function, we have three random variables. The diameter d will be treated as a normal distribution. Its mean and standard deviation can be determined per Equation (1.1). The distribution parameters of these three random variables are listed in Table 2.38.

Table 2.38: The distribution parameters of random variables in Equation (e)

K (log-normal)		F_a (klb)		d (in)	
$\mu_{\ln K}$	$\sigma_{\ln K}$	μ_{Fa}	σ_{Fa}	μ_d	σ_d
41.738	0.357	14.11	1.51	0.85	0.00125

(3) Reliability of the bar.

The limit state function (e) contains two normal distributions and one log-normal distribution. We will use the R-F method to calculate its reliability, which is displayed in Appendix A.2. We can follow the procedure and the flowchart of the R-F method to create a MATLAB program. The iterative results are listed in Table 2.39. From the iterative results, the reliability index β and the corresponding reliability R of the bar in this example are:

$$\beta = 2.172022 \qquad R = \Phi(2.172022) = 0.9851.$$

∎

Table 2.39: The iterative results of Example 2.22 by the R-F method

Iterative #	K^*	F_a^*	d^*	β^*	$|\Delta\beta^*|$
1	1.43E+18	14.11	0.775131	2.161288	
2	9.22E+17	17.13139	0.852768	2.17213	0.010842
3	9.5E+17	17.04937	0.84992	2.172022	0.000108
4	9.51E+17	17.05197	0.849898	2.172022	1.01E-07

Example 2.23

A bar with a diameter $0.820 \pm 0.005''$ is subjected to model #6 cyclic axial loading spectrum listed in Table 2.40. The ultimate material strength S_u is 75 (ksi). Three parameters of the component fatigue strength index K are $m = 8.21$, $\mu_{\ln K} = 41.738$, and $\sigma_{\ln K} = 0.357$. For the component fatigue strength index K, the stress unit is ksi. Calculate the reliability of this bar.

Table 2.40: Model #6 cyclic axial loading spectrum for Example 2.23

Level #	The Number of Cycles n_{Li} (constant)	The Fully Reversed Axial Loading Amplitude F_{ai} (klb) (normal distribution)	
		$\mu_{F_{ai}}$	$\sigma_{F_{ai}}$
1	5,000	22.15	3.25
2	200,000	12.45	1.5

Solution:

(1) The cyclic axial stress and the component fatigue damage index.

Since the cyclic axial loadings are fully reversed cyclic axial loading, the fully reversed cyclic axial stress amplitudes will be as follows.

In the level #1, the fully reversed axial stress amplitude σ_{a1} is

$$\sigma_{a1} = \frac{F_{a1}}{\pi d^2/4} = \frac{4F_{a1}}{\pi d^2}. \tag{a}$$

In the level #2, the fully reversed axial stress amplitude σ_{a2} is

$$\sigma_{a1} = \frac{F_{a2}}{\pi d^2/4} = \frac{4F_{a2}}{\pi d^2}. \tag{b}$$

The component fatigue damage index D of the bar under this model #6 cyclic loading per Equation (2.85) is:

$$D = n_{L1}\left(\frac{4F_{a1}}{\pi d^2}\right)^m + n_{L1}\left(\frac{4F_{a2}}{\pi d^2}\right)^m. \tag{c}$$

(2) The limit state function of this bar.

The limit state function of a bar due to model #6 cyclic axial loading spectrum can be established per Equation (2.88):

$$g(K, F_{a1}, F_{a2}, d) = K - n_{L1}\left(\frac{4F_{a1}}{\pi d^2}\right)^m - n_{L1}\left(\frac{4F_{a2}}{\pi d^2}\right)^m = \begin{cases} > 0 & \text{Safe} \\ 0 & \text{Limit state} \\ < 0 & \text{Failure.} \end{cases} \tag{d}$$

The diameter d will be treated as a normal distribution. Its mean and standard deviation can be determined per Equation (1.1). There are four random variables in the limit state function (d). K is a log-normal distribution. F_{a1}, F_{a2}, and d are normal distributions. Their distribution parameters in Equation (d) are listed in Table 2.41.

Table 2.41: The distribution parameters of random variables in Equation (d)

K (log-normal)		F_{a1} (klb)		F_{a2} (klb)		d (in)	
$\mu_{\ln K}$	σ_{F_a}	μ_{Fa1}	σ_{Fa1}	μ_{Fa2}	σ_{Fa2}	μ_d	σ_d
41.738	0.357	22.15	3.25	12.45	1.5	0.82	0.00125

(3) Reliability of the bar.

We will use the Monte Carlo method to calculate the reliability of this example. The Monte Carlo method is displayed in Appendix A.3. We can follow the Monte Carlo method and the program flowchart to create a MATLAB program.

Since the limit state function is not too complicated, we will use the trial number $N = 1,598,400$. The reliability of this component R by the Monte Carlo method is

$$R = \frac{1,578,582}{1,598,400} = 0.9876.$$

■

2.9.7 RELIABILITY OF A COMPONENT UNDER CYCLIC DIRECT SHEARING LOADING

Per Equation (2.87) or Equation (2.88), we can establish the limit state function of a component under any type of cyclic direct shearing loading spectrum and then calculate its reliability. In this section, we will use two examples to demonstrate how to calculate the reliability of a component under a cyclic direct-shearing loading spectrum.

Example 2.24
A single-shearing pin with a diameter $1.125 \pm 0.005''$ is under a zero-to-maximum cyclic direct shearing loading. The maximum shear loading $V_{\max}$ of this cyclic shearing loading can be treated as a constant $V_{\max} = 26.75$ (klb). The number of cycles n_L of this cyclic shearing loading is also treated as a constant $n_L = 500,000$ (cycles). The ultimate material strength S_u of the pin is 75 (ksi). Three parameters of the component fatigue strength index K on the critical section for the cyclic shear loading are $m = 8.21$, $\mu_{\ln K} = 37.308$, and $\sigma_{\ln K} = 0.518$. For the component fatigue strength index K, the stress unit is ksi. Calculate the reliability of the pin.

Solution:

(1) The cyclic direct shearing stress and the component fatigue damage index.

The mean shear stress τ_m and the shear stress amplitude τ_a of the pin due to this zero-to-maximum cyclic shearing loading are:

$$\tau_m = \frac{V_m}{A} = \frac{V_{\max}/2}{\pi d^2/4} = \frac{2V_{\max}}{\pi d^2} \tag{a}$$

$$\tau_a = \frac{V_a}{A} = \frac{V_{\max}/2}{\pi d^2/4} = \frac{2V_{\max}}{\pi d^2}. \tag{b}$$

Since this is non-zero-mean cyclic shear stress, the equivalent stress amplitude of a fully reversed cyclic shear stress is:

$$\tau_{a-eq} = \tau_a \frac{S_u}{(S_u - \tau_m)} = \frac{2V_{\max}}{\pi d^2} \frac{S_u}{(S_u - 2V_{\max}/\pi d^2)} = \frac{2V_{\max}S_u}{(\pi d^2 S_u - 2V_{\max})}. \tag{c}$$

The component fatigue damage index of this pin under model #1 cyclic shear stress per Equation (2.84) is:

$$D = n_L \left[\frac{2V_{\max}S_u}{(\pi d^2 S_u - 2V_{\max})} \right]^{8.21}. \tag{d}$$

(2) The limit state function.

The limit state function of the pin under model #1 cyclic loading spectrum per Equation (2.87) is:

$$g(K, d) = K - n_L \left[\frac{2V_{\max}S_u}{(\pi d^2 S_u - 2V_{\max})} \right]^{8.21} = \begin{cases} > 0 & \text{Safe} \\ 0 & \text{Limit state} \\ < 0 & \text{Failure.} \end{cases} \tag{e}$$

There are two random variables in the limit state function (e). The dimension d can be treated as a normal distribution, and its mean and its standard deviation can be calculated per Equation (1.1). The distribution parameters in the limit state function (e) are listed in Table 2.42.

(3) The reliability of the single-shear pin.

We will use the Monte Carlo method to calculate the reliability of this example. We can follow the Monte Carlo method and the program flowchart in Appendix A.3 to create a MATLAB program. Since the limit state function is not too complicated, we will use the trial number $N = 1{,}598{,}400$. The reliability of this component R by the Monte Carlo method is

$$R = \frac{1{,}583{,}621}{1{,}598{,}400} = 0.9908.$$

Table 2.42: The distribution parameters of random variables in Equation (e)

K (lognormal)		d (in)	
$\mu_{\ln K}$	$\sigma_{\ln K}$	μ_d	σ_d
37.308	0.518	1.125	0.00125

■

Example 2.25
A double-shearing pin with a diameter $0.500 \pm 0.005''$ is subjected to model #3 cyclic shear loading spectrum, as shown in Table 2.43. The ultimate material strength S_u of the pin is 75 (ksi). Three parameters of the component fatigue strength index K on the critical section for the cyclic shear loading are $m = 8.21$, $\mu_{\ln K} = 37.308$, and $\sigma_{\ln K} = 0.518$. For the component fatigue strength index K, the stress unit is ksi. Calculate the reliability of the pin.

Table 2.43: Model #3 cyclic shearing loading spectrum for Example 2.25

Number of Cycles n_L	Mean of the Cyclic Shear Loading V_m (klb)	Amplitude of Cyclic Shear Loading V_a (klb) (normal distribution)	
		μ_{V_a}	σ_{V_a}
600,000	3.422	4.815	0.6

Solution:

(1) The cyclic shearing stress and the component fatigue damage index.

The mean shear stress τ_m and the shear stress amplitude τ_a of the pin due to this cyclic shearing loading are:

$$\tau_m = \frac{V_m/2}{A} = \frac{V_m/2}{\pi d^2/4} = \frac{2V_m}{\pi d^2} \tag{a}$$

$$\tau_a = \frac{V_a/2}{A} = \frac{V_a/2}{\pi d^2/4} = \frac{2V_a}{\pi d^2}. \tag{b}$$

Since this is non-zero mean cyclic shear stress, the equivalent stress amplitude of a fully reversed cyclic shear stress is:

$$\tau_{a-eq} = \tau_a \frac{S_u}{(S_u - \tau_m)} = \frac{2V_a S_u}{(\pi d^2 S_u - 2V_m)}. \tag{c}$$

The component fatigue damage index of this pin under model #3 cyclic shear loading spectrum per Equation (2.84) is:

$$D = n_L \left[\frac{2V_a S_u}{(\pi d^2 S_u - 2V_m)} \right]^{8.21}.$$

(d)

(2) The limit state function.

The limit state function of the pin under model #3 cyclic shearing loading spectrum per Equation (2.87) is:

$$g(K, V_a, d) = K - n_L \left[\frac{2V_a S_u}{(\pi d^2 S_u - 2V_m)} \right]^{8.21} = \begin{cases} > 0 & \text{Safe} \\ 0 & \text{Limit state} \\ < 0 & \text{Failure.} \end{cases}$$

(e)

There are three random variables in the limit state function (e). The dimension d can be treated as a normal distribution, and its mean and standard deviation can be calculated per Equation (1.1). The distribution parameters in the limit state function (e) are listed in Table 2.44.

(3) The reliability of the double-shear pin.

We will use the Monte Carlo method to calculate the reliability of this example. We can follow the Monte Carlo method and the program flowchart in Appendix A.3 to create a MATLAB program. Since the limit state function is not too complicated, we will use the trial number $N = 1,598,400$. The reliability of this component R by the Monte Carlo method is

$$R = \frac{1,581,583}{1,598,400} = 0.9895.$$

∎

Table 2.44: The distribution parameters of random variables in Equation (e)

K (lognormal)		V_a (klb)		d (in)	
$\mu_{\ln K}$	$\sigma_{\ln K}$	μ_{Va}	σ_{Va}	μ_d	σ_d
37.308	0.518	4.815	0.6	0.5	0.00125

2.9.8   RELIABILITY OF A SHAFT UNDER CYCLIC TORSION LOADING

Per Equation (2.87) or Equation (2.88), we can establish the limit state function of a shaft under any type of cyclic torsion loading spectrum and then calculate its reliability by using the H-L method, or the R-F method and the Monte Carlo method. In this section, we will use two examples to demonstrate how to calculate the reliability of a component under cyclic torsion loading spectrum.

Example 2.26
A shaft with a diameter $1.250 \pm 0.005''$ is subjected to model #4 cyclic torsion loading spectrum as listed in Table 2.45. The ultimate material strength S_u of the shaft is 75 (ksi). Three parameters of the component fatigue strength index K on the critical section for the cyclic torsion loading are $m = 8.21$, $\mu_{\ln K} = 37.308$ and $\sigma_{\ln K} = 0.518$. For the component fatigue strength index K, the stress unit is ksi. Calculate the reliability of the shaft.

Table 2.45: The model #4 cyclic torsion loading spectrum for Example 2.26

Loading Level #	Number of Cycles n_{Li}	Mean T_{mi} of the Cyclic Torque (klb.in)	Amplitude T_{ai} of the Cyclic Torque (klb.in)
1	6,000	2.25	5.13
2	500,000	2.25	9.42

Solution:

(1) The cyclic torsion stress and the component fatigue damage index.

For the loading level #1, we have the mean shear stress τ_{m1}, the shear stress amplitude τ_{a1} and their corresponding equivalent shear stress amplitude τ_{eq-1}

$$\tau_{m1} = \frac{T_{m1} \times d/2}{J} = \frac{T_{m1} \times d/2}{\pi d^4/32} = \frac{16 T_{m1}}{\pi d^3} \tag{a}$$

$$\tau_{a1} = \frac{T_{a1} \times d/2}{J} = \frac{T_{a1} \times d/2}{\pi d^4/32} = \frac{16 T_{a1}}{\pi d^3} \tag{b}$$

$$\tau_{eq-1} = \tau_{a1} \frac{S_u}{S_u - \tau_{m1}} = \frac{16 \tau_{a1} S_u}{\pi d^3 S_u - 16 T_{m1}}. \tag{c}$$

For the loading level #2, by repeating the above calculation, we have

$$\tau_{eq-2} = \tau_{a2} \frac{S_u}{S_u - \tau_{m2}} = \frac{16 \tau_{a2} S_u}{\pi d^3 S_u - 16 T_{m2}}. \tag{d}$$

The component fatigue damage index of this shaft under model #4 cyclic torsion stress per Equation (2.85) is:

$$D = n_{L1} \left[\frac{16 \tau_{a1} S_u}{\pi d^3 S_u - 16 T_{m1}} \right]^{8.21} + n_{L2} \left[\frac{16 \tau_{a2} S_u}{\pi d^3 S_u - 16 T_{m2}} \right]^{8.21}. \tag{e}$$

(2) The limit state function of the shaft.

The limit state function of the shaft under model #4 cyclic loading spectrum per Equation (2.88) is :

$$g\,(K,d) = K - n_{L1}\left[\frac{16\tau_{a1}S_u}{\pi d^3 S_u - 16T_{m1}}\right]^{8.21} - n_{L2}\left[\frac{16\tau_{a2}S_u}{\pi d^3 S_u - 16T_{m2}}\right]^{8.21}$$

$$= \begin{cases} > 0 & \text{Safe} \\ 0 & \text{Limit state} \\ < 0 & \text{Failure.} \end{cases} \tag{f}$$

There are two random variables in the limit state function (f). The dimension d can be treated as a normal distribution. Its mean and standard deviation can be calculated per Equation (1.1). The distribution parameters in the limit state function (f) are listed in Table 2.46.

(3) The reliability of the shaft.

We will follow the Monte Carlo method and the program flowchart in Appendix A.3 to create a MATLAB program. Since the limit state function is not too complicated, we will use the trial number $N = 1,598,400$. The reliability of this shaft R by the Monte Carlo method is

$$R = \frac{1,582,682}{1,598,400} = 0.9902.$$

∎

Table 2.46: The distribution parameters of random variables in Equation (f)

K (lognormal)		d (in)	
$\mu_{\ln K}$	$\sigma_{\ln K}$	μ_d	σ_d
37.308	0.518	1.25	0.00125

Example 2.27
A shaft with a diameter $1.500 \pm 0.005''$ is subjected to model #3 cyclic torsion loading spectrum as listed in Table 2.47. The ultimate material strength S_u of the shaft is 75 (ksi). Three parameters of the component fatigue strength index K on the critical section for the cyclic shear loading are $m = 8.21$, $\mu_{\ln K} = 37.308$ and $\sigma_{\ln K} = 0.518$. For the component fatigue strength index K, the stress unit is ksi. Calculate the reliability of the shaft.

Solution:

(1) The cyclic torsion stress and the component fatigue damage index.

Table 2.47: The model #3 cyclic loading spectrum for Example 2.27

Number of Cycles n_L	Mean Torque T_m (klb.in)	Normally Distributed Torque Amplitude T_a (klb.in)	
		The Mean μ_{T_a}	The Standard Deviation σ_{T_a}
400,000	4.5	8.9	0.85

The mean shear stress τ_m, the shear stress amplitude τ_a and their corresponding equivalent shear stress amplitude τ_{eq} of the shaft due to the model #3 cyclic torque loading are:

$$\tau_m = \frac{T_m \times d/2}{J} = \frac{T_m \times d/2}{\pi d^4/32} = \frac{16T_m}{\pi d^3} \tag{a}$$

$$\tau_a = \frac{T_a \times d/2}{J} = \frac{T_a \times d/2}{\pi d^4/32} = \frac{16T_a}{\pi d^3} \tag{b}$$

$$\tau_{eq} = \tau_a \frac{S_{us}}{S_{us} - \tau_m} = \frac{16T_a S_{us}}{\pi d^3 S_{us} - 16T_m}. \tag{c}$$

The component fatigue damage index of this shaft under model #3 cyclic shear stress per Equation (2.84) is:

$$D = n_L \left[\frac{16\tau_a S_u}{\pi d^3 S_u - 16T_m} \right]^{8.21}. \tag{d}$$

(2) The limit state function of this shaft.

The limit state function of the shaft under model #3 cyclic loading spectrum per Equation (2.87) is:

$$g(K, T_a, d) = K - n_L \left[\frac{16T_a S_u}{\pi d^3 S_u - 16T_m} \right]^{8.21} = \begin{cases} > 0 & \text{Safe} \\ 0 & \text{Limit state} \\ < 0 & \text{Failure.} \end{cases} \tag{e}$$

There are three random variables in the limit state function (e). The dimension d can be treated as a normal distribution, and its mean and standard deviation can be calculated per Equation (1.1). The distribution parameters in the limit state function (e) are listed in Table 2.48.

(3) The reliability of the shaft.

The limit state function (e) contains two normal distributions and one log-normal distribution. We will follow the procedure of the R-F method in Section 3.2.4 and the flowchart in Appendix A.2 to create a MATLAB program. The iterative results are listed in Table 2.49.

From the iterative results, the reliability index β and corresponding reliability R of the shaft in this example are:

$$\beta = 2.278915 \qquad R = \Phi(2.278915) = 0.9887.$$

∎

Table 2.48: The distribution parameters of random variables in Equation (e)

K (lognormal)		T_a (klb.in)		d (in)	
$\mu_{\ln K}$	$\sigma_{\ln K}$	μ_{T_a}	σ_{T_a}	μ_d	σ_d
37.308	0.518	8.9	0.85	1.500	0.00125

Table 2.49: The iterative results of Example 2.27 by the R-F method

Iterative #	K^*	T_a^*	d^*	β^*	$|\Delta\beta^*|$
1	1.82E+16	1.5	11.57999	2.220042	
2	2.13E+15	1.499878	8.913915	2.158924	0.061117
3	5.11E+15	1.499899	9.915022	2.265797	0.106873
4	7.37E+15	1.499887	10.36785	2.278682	0.012884
5	7.76E+15	1.499883	10.43223	2.278909	0.000228
6	7.75E+15	1.499882	10.43026	2.278915	5.31E-06

2.9.9   RELIABILITY OF A BEAM UNDER CYCLIC BENDING LOADING

Per Equation (2.87) or Equation (2.88), we can establish the limit state function of a beam under any type of cyclic bending loading spectrum and then calculate its reliability. In this section, we will use two examples to demonstrate how to calculate the reliability of a beam under cyclic bending loading spectrum.

Example 2.28

The critical section of a beam with a rectangular cross-section is subjected to model #5 cyclic bending loading spectrum listed in Table 2.50. The height h and the width b of the critical cross-section are $h = 2.500 \pm 0.010''$ and $b = 4.00 \pm 0.010''$. The ultimate material strength S_u of the beam is 61.5 (ksi). Three parameters of the component fatigue strength index K on the critical section for the cyclic bending loading are $m = 6.38$, $\mu_{\ln K} = 32.476$, and $\sigma_{\ln K} = 0.279$. For the component fatigue strength index K, the stress unit is ksi. Calculate the reliability of the beam.

Table 2.50: Model #5 cyclic bending loading spectrum for Example 2.28

Stress Level #i	Cyclic Bending Moment (klb.in)		Number of Cycles N_{Li} (normal distribution)	
	M_{mi}	M_{ai}	$\mu_{N_{Li}}$	$\sigma_{N_{Li}}$
1	80.2	61.55	250,000	3,000
2	80.2	108.17	5,000	450

Solution:

(1) The cyclic bending stress and the component fatigue damage index.

For the stress level #1, the mean bending stress σ_{m1}, the bending stress amplitude σ_{a1}, and its corresponding equivalent bending stress amplitude σ_{eq1} are

$$\sigma_{m1} = \frac{M_{m1} \times h/2}{I} = \frac{M_{m1} \times h/2}{bh^3/12} = \frac{6M_{m1}}{bh^2} \tag{a}$$

$$\sigma_{a1} = \frac{M_{a1} \times h/2}{I} = \frac{M_{a1} \times h/2}{bh^3/12} = \frac{6M_{a1}}{bh^2} \tag{b}$$

$$\sigma_{eq1} = \sigma_{a1} \frac{S_u}{S_u - \sigma_{m1}} = \frac{6M_{a1}S_u}{bh^2 S_u - 6M_{m1}}. \tag{c}$$

For the stress level #2, by repeating (a), (b), and (c), we have

$$\sigma_{eq2} = \sigma_{a2} \frac{S_u}{S_u - \sigma_{m2}} = \frac{6M_{a1}S_u}{bh^2 S_u - 6M_{m2}}. \tag{d}$$

The component fatigue damage index of this beam under model #5 cyclic bending stress per Equation (2.85) is:

$$D = N_{L1} \left[\frac{6M_{a1}S_u}{bh^2 S_u - 6M_{m1}} \right]^{6.38} + N_{L2} \left[\frac{6M_{a2}S_u}{bh^2 S_u - 6M_{m2}} \right]^{6.38}. \tag{e}$$

(2) The limit state function of the beam.

The limit state function of the beam under model #5 cyclic loading spectrum per Equation (2.87) is:

$$g(K, N_{L1}, N_{L2}, b, h) = K - N_{L1} \left[\frac{6M_{a1}S_u}{bh^2 S_u - 6M_{m1}} \right]^{6.38} - N_{L2} \left[\frac{6M_{a2}S_u}{bh^2 S_u - 6M_{m2}} \right]^{6.38}$$

$$= \begin{cases} > 0 & \text{Safe} \\ 0 & \text{Limit state} \\ < 0 & \text{Failure.} \end{cases} \tag{f}$$

There are five random variables in the limit state function (f). The dimensions h and b can be treated as normal distributions. Their means and standard deviations can be calculated per Equation (1.1). The distribution parameters in the limit state function (f) are listed in Table 2.51.

(3) The reliability of the beam.

We will follow the Monte Carlo method and the program flowchart in Appendix A.3 to create a MATLAB program. Since the limit state function is not too complicated, we will use the trial number $N = 1{,}598{,}400$. The reliability R of this beam by the Monte Carlo method is

$$R = \frac{1{,}514{,}162}{1{,}598{,}400} = 0.9473.$$

∎

Table 2.51: The distribution parameters of random variables in Equation (e)

K (lognormal)		N_{L1} (normal)		N_{L2} (normal)		b (in)		h (in)	
$\mu_{\ln K}$	$\sigma_{\ln K}$	$\mu_{N_{L1}}$	$\sigma_{N_{L1}}$	$\mu_{N_{L2}}$	$\sigma_{N_{L2}}$	μ_b	σ_b	μ_h	σ_h
32.476	0.279	250,000	3,000	5,000	450	4.000	0.00125	2.500	0.00125

Example 2.29
The critical section of a beam with a circular cross-section is subjected to model #2 cyclic bending spectrum listed in Table 2.52. The diameter d of the critical cross-section is $d = 2.500 \pm 0.010''$. The ultimate material strength S_u of the beam is 61.5 (ksi). Three parameters of the component fatigue strength index K on the critical section for the cyclic shear loading are $m = 6.38$, $\mu_{\ln K} = 32.476$, and $\sigma_{\ln K} = 0.279$. For the component fatigue strength index K, the stress unit is ksi. Calculate the reliability of the beam.

Table 2.52: The model #2 cyclic bending loading spectrum for Example 2.29

Number of Cycles	Mean Bending Moment M_m (klb.in)	Normally Distributed Bending Moment Amplitude M_a (klb.in)	
		μ_{M_a}	σ_{M_a}
550,000	28.2	16.8	3.19

Solution:

(1) The cyclic bending stress and the component fatigue damage index.

The mean bending stress σ_m, the bending stress amplitude σ_a, and its corresponding equivalent bending stress amplitude σ_{eq} of the beam due to model #2 cyclic bending loading are:

$$\sigma_m = \frac{M_m \times d/2}{I} = \frac{M_m \times d/2}{\pi d^4/64} = \frac{32 M_m}{\pi d^3} \tag{a}$$

$$\sigma_a = \frac{M_a \times d/2}{I} = \frac{M_a \times d/2}{\pi d^4/64} = \frac{32 M_a}{\pi d^3} \tag{b}$$

$$\sigma_{eq} = \sigma_a \frac{S_u}{S_u - \sigma_m} = \frac{32 M_a S_u}{\pi d^3 S_u - 32 M_m}. \tag{c}$$

The component fatigue damage index of this beam under model #2 cyclic bending stress per Equation (2.84) is:

$$D = n_L \left[\frac{32 M_a S_u}{\pi d^3 S_u - 32 M_m} \right]^{6.38}. \tag{d}$$

(2) The limit state function of the beam.

The limit state function of the beam under model #2 cyclic bending loading spectrum per Equation (2.87) is:

$$g(K, M_a, d) = K - n_L \left[\frac{32 M_a S_u}{\pi d^3 S_u - 32 M_m} \right]^{6.38} = \begin{cases} > 0 & \text{Safe} \\ 0 & \text{Limit state} \\ < 0 & \text{Failure.} \end{cases} \tag{e}$$

There are three random variables in the limit state function (e). The dimension d can be treated as a normal distribution. Its mean and standard deviation can be calculated per Equation (1.1). The distribution parameters in the limit state function (e) are listed in Table 2.53.

(3) The reliability of the beam.

The limit state function (e) contains two normal distributions and one lognormal distribution. We will use the R-F method to calculate its reliability. We can follow the procedure and the flowchart in Appendix A.2 to create a MATLAB program. The iterative results are listed in Table 2.54. From the iterative results, the reliability index β and corresponding reliability R of the beam in this example are:

$$\beta = 1.624707 \qquad R = \Phi(1.624707) = 0.9479.$$

■

Table 2.53: The distribution parameters of random variables in Equation (e)

K (lognormal)		M_a (klb.in)		d (in)	
$\mu_{\ln K}$	$\sigma_{\ln K}$	μ_{M_a}	σ_{M_a}	μ_d	σ_d
32.476	0.279	16.8	3.19	2.500	0.00125

Table 2.54: The iterative results of Example 2.29 by the R-F method

Iterative #	K^*	M_a^*	d^*	β^*	$|\Delta\beta^*|$
1	1.32E+14	16.8	2.345547	1.557169	
2	1.14E+14	21.63949	2.494604	1.624744	0.067575
3	1.12E+14	21.76716	2.499996	1.624707	3.69E-05

2.9.10 RELIABILITY OF A COMPONENT UNDER CYCLIC COMBINED LOADING

Component under cyclic combined loading is very complicated because the frequencies of independent loadings might not be in phase. In this section, we will only discuss a rotating shaft under a combined-torques-bending loading.

For a rotating shaft, the cyclic bending stress on a rotating shaft is mainly due to the rotation of the shaft under a bending moment. For a combined stress fatigue issue, we can use the Von Mises stress as the equivalent stress to run related fatigue calculation [2]. We will use the following assumptions to study the fatigue issue of a rotating shaft.

(1) The cyclic stress due to the combined loading will be a cyclic Von Mises stress.

(2) The mean of the cyclic Von Mises stress is mainly induced by the acting torque on the rotating shaft.

(3) The stress amplitude of the cyclic Von Mises stress is mainly induced by the acting bending moment on the rotating shaft.

(4) The modified Goodman approach will be used to consider the effect of mean stress in cyclic Von Mises stress.

Now, we will discuss the cyclic combined loading ($n_{Li}, T_i, M_i, i = 1, 2, \ldots, L$), where i is the ith loading level. n_{Li}, T_i, and M_i are the number of cycles, the torque and the bending moment, respectively, in the ith loading level. L is the number of different combined loading levels. n_{Li} could be a constant number or a distributed number of cycles in a combined loading level. M_i can be a constant bending moment or a distributed bending moment in a combined loading level. But T_i will be a constant in each loading level. In the ith combined loading level with a bending moment M_i and a torque T_i, the mean stress σ_{von-mi}, and the stress amplitude σ_{von-ai} of a cyclic Von Mises stress will be:

$$\sigma_{von-mi} = \sqrt{3}K_{fs}\tau_{T_i} \tag{2.89}$$

$$\sigma_{von-ai} = K_f\sigma_{M_i}, \tag{2.90}$$

where K_{fs} and K_f are the fatigue stress concentration factors for torsion stress and bending stress on the critical section, respectively, which has been discussed in Section 2.6. τ_{T_i} is the outer layer nominal shear stress of the rotating shaft on the critical section due to the torque T_i. σ_{M_i} is the maximum nominal bending stress on the rotating shaft due to a bending moment M_i.

Since it is a non-zero-mean cyclic combined stress, we need to convert it into an equivalent fully-reversed cyclic stress. The equivalent stress amplitude σ_{a-eqi} of this transferred fully reversed cyclic stress is:

$$\sigma_{a-eqi} = \sigma_{von-ai}\frac{S_u}{S_u - \sigma_{von-mi}} = \frac{S_u K_f\sigma_{M_i}}{S_u - \sqrt{3}K_{fs}\tau_{T_i}}. \tag{2.91}$$

The component fatigue damage index D will be:

$$D = \sum_{i=1}^{L} n_{Li}\left(\frac{S_u K_f\sigma_{M_i}}{S_u - \sqrt{3}K_{fs}\tau_{T_i}}\right)^m. \tag{2.92}$$

For a solid shaft with a diameter d, Equation (2.92) for the component fatigue damage index D will become

$$D = \sum_{i=1}^{L} n_{Li}\left(\frac{32S_u K_f M_i}{\pi d^3 S_u - 16\sqrt{3}K_{fs}T_i}\right)^m. \tag{2.93}$$

For a hollow solid shaft with an inner diameter d_0 and an outer diameter d, Equation (2.92) for the component fatigue damage index D will become

$$D = \sum_{i=1}^{L} n_{Li}\left[\frac{32S_u K_f M_i d}{\pi\left(d^4 - d_0^4\right)S_u - 16\sqrt{3}K_{fs}T_i d}\right]^m. \tag{2.94}$$

The component fatigue strength K for a cyclic combined loading will be obtained based on the cyclic bending stress because the cyclic Von Mises stress is mainly due to the bending moment and the shaft rotation. Therefore, the limit state function of a rotating shaft due to a combined loading is:

$$g(K, D) = K - D = \begin{cases} K - \sum_{i=1}^{L} n_{Li}\left[\dfrac{32S_u K_f M_i}{\pi d^3 S_u - 16\sqrt{3}K_{fs}T_i}\right]^m & \text{solid shaft} \\[4mm] K - \sum_{i=1}^{L} n_{Li}\left[\dfrac{32S_u K_f M_i d}{\pi\left(d^4 - d_0^4\right)S_u - 16\sqrt{3}K_{fs}T_i d}\right]^m & \text{hollow shaft.} \end{cases} \tag{2.95}$$

Equation (2.95) can be used to calculate the reliability of a rotating shaft under cyclic combined loading. Now, we will use two examples to demonstrate how to calculate the reliability of a rotating shaft under cyclic combined loading.

Example 2.30

The critical section of a solid rotating shaft with a diameter $1.750 \pm 0.005''$ is subjected to cyclic combined loading listed in Table 2.55. The ultimate material strength S_u of the beam is 75 (ksi). Three parameters of the component fatigue strength index K on the critical section for the cyclic bending loading are $m = 8.21$, $\mu_{\ln K} = 41.738$, and $\sigma_{\ln K} = 0.357$. For the component fatigue strength index K, the stress unit is ksi. Calculate the reliability of the shaft.

Table 2.55: The cyclic combined loading spectrum for Example 2.30

Stress Level # i	Number of Cycles n_{Li}	Torque T_i (klb.in)	Bending Moment M_i (klb.in)
1	5,500	17.75	15.75
2	580,000	10.29	10.15

Solution:

(1) The cyclic Von Mises stress and the component fatigue damage index.

For the stress level #1, the mean Von Mises stress σ_{von-m1} per Equation (2.49), the Von Mises stress amplitude σ_{von-a1} per Equation (2.90) and its corresponding equivalent Von Mises stress amplitude σ_{a-eq1} per Equation (2.91) are

$$\sigma_{von-m1} = \sqrt{3}K_{fs}\tau_{T_1} = \sqrt{3}\frac{16T_1}{\pi d^3} \tag{a}$$

$$\sigma_{von-a1} = K_f\sigma_{M_1} = \frac{32M_1}{\pi d^3} \tag{b}$$

$$\sigma_{a-eq1} = \sigma_{von-a1}\frac{S_u}{S_u - \sigma_{von-m}} = \frac{32M_1 S_u}{\pi d^3 S_u - 16\sqrt{3}T_1}. \tag{c}$$

For the stress level #2, we repeat the above calculations,

$$\sigma_{a-eq2} = \frac{32M_2 S_u}{\pi d^3 S_u - 16\sqrt{3}T_2}. \tag{d}$$

The component fatigue damage index of this shaft under the cyclic combined loading per Equation (2.94) is:

$$D = n_{L1}\left[\frac{32M_1 S_u}{\pi d^3 S_u - 16\sqrt{3}T_1}\right]^{8.21} + n_{L2}\left[\frac{32M_2 S_u}{\pi d^3 S_u - 16\sqrt{3}T_2}\right]^{8.21}. \tag{e}$$

(2) The limit state function of the shaft.

The limit state function of the rotating shaft in this example per Equation (2.95) is:

$$g\left(K, d\right) = K - n_{L1} \left[\frac{32 M_1 S_u}{\pi d^3 S_u - 16\sqrt{3} T_1}\right]^{8.21} - n_{L2} \left[\frac{32 M_2 S_u}{\pi d^3 S_u - 16\sqrt{3} T_2}\right]^{8.21}$$

$$= \begin{cases} > 0 & \text{Safe} \\ 0 & \text{Limit state} \\ < 0 & \text{Failure.} \end{cases} \qquad \text{(f)}$$

There are two random variables in the limit state function (f). The dimension d can be treated as a normal distribution. Its mean and standard deviation can be calculated per Equation (1.1). The distribution parameters in the limit state function (f) are listed in Table 2.56.

(3) The reliability of the rotating shaft.

We will follow the Monte Carlo method and the program flowchart in Appendix A.3 to create a MATLAB program. Since the limit state function is not too complicated, we will use the trial number $N = 1,598,400$. The reliability of this shaft R by the Monte Carlo method is

$$R = \frac{1,581,687}{1,598,400} = 0.9895.$$

∎

Table 2.56: The distribution parameters of random variables in Equation (e)

K (lognormal)		d (in)	
$\mu_{\ln K}$	$\sigma_{\ln K}$	μ_d	σ_d
41.738	0.357	1.750	0.00125

Example 2.31

The critical section of a solid rotating shaft with a diameter $2.150 \pm 0.005''$ is subjected to cyclic combined loading listed in Table 2.57. The ultimate material strength S_u of the beam is 75 (ksi). Three parameters of the component fatigue strength index K on the critical section for the cyclic bending loading are $m = 8.21$, $\mu_{\ln K} = 41.738$, and $\sigma_{\ln K} = 0.357$. For the component fatigue strength index K, the stress unit is ksi. Calculate the reliability of the rotating shaft.

Solution:

1. The cyclic Von Mises stress and the component fatigue damage index.

Table 2.57: The cyclic combined loading spectrum for Example 2.31

Number of Cycles	Torque T (klb.in)	Normally Distributed Bending Moment M (klb.in)	
		μ_M	σ_M
450,000	21.15	21.34	1.31

The mean Von Mises stress σ_{von-m} per Equation (2.49), the Von Mises stress amplitude σ_{von-a} per Equation (2.90), and its corresponding equivalent Von Mises stress amplitude σ_{a-eq} per Equation (2.91) of the rotating shaft due to the cyclic combined loadings are:

$$\sigma_{von-m} = \sqrt{3}K_{fs}\tau_T = \sqrt{3}\frac{16T}{\pi d^3} \tag{a}$$

$$\sigma_{von-a} = K_f\sigma_M = \frac{32M}{\pi d^3} \tag{b}$$

$$\sigma_{a-eq} = \sigma_{von-a}\frac{S_u}{S_u - \sigma_{von-m}} = \frac{32M S_u}{\pi d^3 S_u - 16\sqrt{3}T}. \tag{c}$$

The component fatigue damage index of this shaft under the cyclic combined loading per Equation (2.94) is:

$$D = N_L \left[\frac{32M S_u}{\pi d^3 S_u - 16\sqrt{3}T}\right]^{8.21}. \tag{d}$$

(2) The limit state function of the rotating shaft.

The limit state function of the rotating shaft under the cyclic combined loading in this example per Equation (2.95) is:

$$g(K, M, d) = K - N_L \left[\frac{32M S_u}{\pi d^3 S_u - 16\sqrt{3}T}\right]^{8.21} = \begin{cases} > 0 & \text{Safe} \\ 0 & \text{Limit state} \\ < 0 & \text{Failure.} \end{cases} \tag{e}$$

There are three random variables in the limit state function (e). The dimension d can be treated as a normal distribution. Its mean and standard deviation can be calculated per Equation (1.1). The distribution parameters in the limit state function (e) are listed in Table 2.58.

(3) The reliability of the rotating shaft.

The limit state function (e) contains two normal distributions and one log-normal distribution. We will follow the procedure of the R-F method and the flowchart in Appendix A.2

to create a MATLAB program. The iterative results are listed in Table 2.59. From the iterative results, the reliability index β and corresponding reliability R of the shaft in this example are:

$$\beta = 1.705804 \qquad R = \Phi(1.705804) = 0.9560.$$

■

Table 2.58: The distribution parameters of random variables in Equation (e)

K (lognormal)		M (klb.in)		d (in)	
$\mu_{\ln K}$	$\sigma_{\ln K}$	μ_M	σ_M	μ_d	σ_d
41.738	0.357	21.34	1.31	2.150	0.00125

Table 2.59: The iterative results of Example 2.31 by the R-F method

Iterative #	K^*	M^*	d^*	β^*	$|\Delta\beta^*|$
1	1.43E+18	21.34	2.081	1.722323	
2	8.29E+17	23.18013	2.15866	1.706654	0.015669
3	9.18E+17	23.11098	2.150319	1.705806	0.000849
4	9.24E+17	23.11206	2.149932	1.705804	1.81E-06

2.9.11   RELIABILITY OF A COMPONENT WITH THE K-D MODEL BY THE MONTE CARLO METHOD

The Monte Carlo method is displayed in Appendix A.3. In order to use the Monte Carlo method to calculate the reliability of a component, we must establish its limit state function. When the K-D model is used to describe the component fatigue strength index K and the component fatigue damage index D, we can establish limit state functions for every possible cyclic loading case.

When component is under simple cyclic loading, that is, cyclic axial loading, cyclic shearing loading, cyclic torque loading, or cyclic bending loading, we can use Equation (2.87) to establish a limit state function when simple cyclic loading can be described by model #1, #2, and #3 cyclic loading spectrums. We can use Equation (2.88) to establish a limit state function when simple cyclic loading can be described by model #4, #5, and #6 cyclic loading spectrums.

For cyclic combined loading on a rotating shaft, we can use Equation (2.95) to establish a limit state function.

When a limit state function is established, we can use the Monte Carlo method to calculate the reliability of a component under a cyclic loading spectrum. Lots of examples in Section 2.9 such as Examples 2.24, 2.28, and 2.30 use the Monte Carlo method to calculate the reliability of component under a cyclic loading spectrum.

2.9.12　THE COMPARISON OF RESULTS BY THE K-D MODEL WITH THE RESULTS BY THE P-S-N CURVES

Both the P-S-N curves and the K-D model are probabilistic fatigue theory and can be used to describe material fatigue test data, that is, material fatigue strength. Both can be used to calculate the reliability of a component under cyclic loading spectrums.

　　Table 2.11 in Section 2.8.1 list fatigue test results of 195 fatigue tests on the 10 gauge 6060-T6 sheet-type flat test specimen under 5 different cyclic axial stress levels. The P-S-N curves on these set of fatigue data are shown in Table 2.13 and are redisplayed here for convenience. The K-D model on these set of fatigue test data is listed in Table 2.35 and redisplayed here for conveniences. Now, the following three Examples 2.32–2.34 will be used to compare the results from both fatigue theories.

Table 2.13: The P-N distribution at different fatigue strength levels

Axial Stress Level #	σ_a (ksi)	σ_m (ksi)	Equivalent Stress $\sigma_{a\text{-}eqi}$ (ksi)	Lognormal Distributed Fatigue Life	
				$\mu_{\ln N}$	$\sigma_{\ln N}$
1	20.833	20.833	35.126	11.4736	0.238106
2	22.083	22.083	38.832	11.0410	0.227320
3	22.5	22.5	40.139	10.9551	0.242377
4	22.917	22.917	41.485	10.7831	0.225735
5	23.333	23.333	42.871	10.716	0.241955

Table 2.35: K_0 with three distribution parameters

Material: 6061-T6 10-Gauge Sheet	Sample Size: 195	Test Conditions: Cyclic Axial Stress, Milled Machined Sheet-type Flat Specimen	
Slope of the Traditional S-N Curve: m	Log Normally Distributed K_0		
	Log Mean: $\mu_{\ln K_0}$	Log Standard Deviation: $\sigma_{\ln K_0}$	
3.8812	25.3014	0.245451	

Example 2.32

The aluminum 6061-T6 10 Gauge sheet-type flat fatigue specimen, as shown in Figure 2.8 is subjected to cyclic axial stress with a stress amplitude 20.833 (ksi) and mean stress 20.833 (kis). The number of cycles of this cyclic axial loading is 60,000 (cycles). The ultimate tensile strength of this material is $S_u = 51.2$ (ksi). The P-S-N curve of this material is listed in Table 2.13, and the K-D model is listed in Table 2.35. Use the P-S-N curve and the K-D model to calculate its reliability.

Solution:

(1) The P-S-N curve approach.

The P-S-N curve in Table 2.13 shows that the number of cycles to failures N under this cyclic stress follows a lognormal distribution with a mean $\mu_{\ln} = 11.4736$ and the standard deviation $\sigma_{\ln N} = 0.238106$. Since the number of cycles at the specified cyclic stress is a constant, the reliability of this specimen under the specified cyclic stress will be:

$$R = P\,(N \geq n) = P\,(\ln(N) \geq \ln(n)) = \Phi\left(\frac{\mu_{\ln N} - \ln(n)}{\sigma_{\ln N}}\right)$$

$$= \Phi\left(\frac{11.4736 - \ln(60000)}{0.238106}\right) = \Phi(1.98021) = 0.97616,$$

where $\Phi\,(\cdot)$ is the CDF of a standard normal distribution.

(2) The K-D probabilistic model.

The cyclic axial stress is with a stress mean $\sigma_m = 20.833$ (ksi) and the stress amplitude $\sigma_a = 20.833$ (ksi). It is not a fully reversed cyclic stress. It will be converted into an equivalent stress amplitude of a fully reversed cyclic stress per Equation (2.21):

$$\sigma_{\dot{e}q} = \frac{\sigma_a \times S_u}{(S_u - \sigma_m)} = \frac{20.83333 \times 51.2}{(51.2 - 20.83333)} = 35.12623 \text{ (ksi)}.$$

The K-D model of this material under cyclic axial loading is shown in Table 2.35. The fatigue damage index D due to the specified loading per Equation (2.84) will be:

$$D = n\sigma_{eq}^m = 60000 \times (35.12623)^{3.8812} = 5.98489 \times 10^{10}.$$

Since the D in this example is a constant, the reliability of the specimen under the specified cyclic loading will be:

$$R = P\,(K_0 \geq D) = P\,(\ln(K_0) \geq \ln(D)) = \Phi\left(\frac{\mu_{\ln K_0} - \ln(D)}{\sigma_{\ln K_0}}\right)$$

$$= \Phi\left(\frac{25.3014 - 24.8151}{0.238106}\right) = \Phi\,(1.981209) = 0.97622.$$

The results from the P-S-N curve and the K-D model are almost identical. The relative difference is 0.006%.   ∎

Example 2.33

The aluminum 6061-T6 10 Gauge sheet-type fatigue specimen, as shown in Figure 2.8 is subjected to cyclic axial stress with a stress amplitude 20.833 (ksi), and mean stress 20.833 (ksi). The number of cycles n_L of this cyclic axial stress follows a lognormal distribution with a mean $\mu_{\ln n_L} = 10.5$ and a standard deviation $\sigma_{\ln n_L} = 0.35$. The ultimate tensile strength of this material is $S_u = 51.2$ (ksi). The P-S-N curve of this material is listed in Table 2.13, and the K-D model is listed in Table 2.35. Use the P-S-N curve and the K-D probabilistic model to calculate its reliability.

Solution:

(1) The P-S-N curve approach.

Since both the fatigue life N and the number of cycles n_L at the cyclic axial stress level are log-normal distributions, the reliability of this specimen under the specified cyclic axial stress will be:

$$R = P(N \geq n) = P(\ln(N) \geq \ln(n)) = P(\ln(N) - \ln(n) \geq 0)$$

$$= \Phi\left(\frac{\mu_{\ln N} - \mu_{\ln N}}{\sqrt{(\sigma_{\ln N})^2 + (\sigma_{\ln n})^2}}\right) = \Phi\left(\frac{11.4736 - 10.5}{\sqrt{(0.238016)^2 + (0.35)^2}}\right)$$

$$= \Phi(2.29995) = 0.98927.$$

(2) The K-D probabilistic model.

The equivalent stress amplitude of a completely reversed cyclic stress per Equation (2.21) is

$$\sigma_{eq} = \frac{\sigma_a \times S_u}{(S_u - \sigma_m)} = \frac{20.83333 \times 51.2}{(51.2 - 20.83333)} = 35.126 \text{ (ksi)}. \tag{a}$$

The fatigue damage index D due to the specified cyclic axial stress per Equation (2.84) will be:

$$D = n_L \sigma_{eq}^m = n_L \times (35.126)^{3.8812}. \tag{b}$$

This equation can be expressed as:

$$\ln(D) = \ln(n_L) + m \ln(\sigma_{eq}) = \ln(n_L) + 13.812989. \tag{c}$$

Since n_L follows a lognormal distribution, D will also follow a lognormal distribution according to the above equation. The mean and standard deviation of the lognormal distributed D will be:

$$\mu_{\ln D} = \mu_{\ln n_L} + 13.812989 = 10.5 + 13.812989 = 24.312989 \tag{d}$$

$$\sigma_{\ln D} = \sigma_{\ln n_L} = 0.35. \tag{e}$$

Since both the specimen fatigue strength index K_0 and the specimen fatigue damage index D are log-normal distributions, the reliability of this specimen under the specified cyclic axial stress will be:

$$R = P\ (K_0 \geq D) = P\ (\ln K_0 \geq \ln D) = P\ (\ln K_0 - \ln D \geq 0)$$

$$= \Phi \left(\frac{(\mu_{\ln K_0} - \mu_{\ln D})}{\sqrt{(\sigma_{\ln K_0})^2 + (\sigma_{\ln D})^2}} \right)$$

$$= \Phi \left(\frac{(25.3014 - 24.312989)}{\sqrt{(0.238106)^2 + (0.35)^2}} \right)$$

$$= \Phi\ (2.312136) = 0.98962.$$

The results from the P-S-N curve and the K-D model are almost identical. The relative difference is 0.035%. ∎

Example 2.34

The aluminum 6061-T6 10 Gauge sheet-type fatigue specimen, as shown in Figure 2.8 is subjected to two levels of cyclic axial stress listed in Table 2.60. The ultimate tensile strength of this material is $S_u = 51.2$ (ksi). The P-S-N curve of this material is listed in Table 2.13, and the K-D model is listed in Table 2.35. Use the P-S-N curve and the K-D model to calculate its reliability.

Table 2.60: Two levels of axial cyclic stress for Example 2.34

Level #	σ_{ai} (ksi)	σ_{mi} (ksi)	Number of Cycles n_{Li}
1	20.833	20.833	30,000
2	22.5	22.5	15,000

Solution:

(1) The P-S-N curve approach.

This cyclic loading is model #4 cyclic loading spectrum. With the concept of equivalent damage approach discussed in Section 2.8.6, the fatigue damage due to a cyclic stress level #1 can be transferred to stress level #2 with an equivalent cyclic number per Equations (2.45), (2.47), and (2.48):

$$\beta_1 = \frac{\mu_{\ln N1} - \ln (n_{L1})}{\sigma_{\ln N1}} = \frac{11.4736 - \ln(30,000)}{0.238106} = 4.89130. \tag{a}$$

With the same fatigue damage, that is, the same reliability index β, the equivalent number of cycles n_{1-2eq} of the stress level #1 into the stress level #2 is:

$$n_{eq1-2} = \exp\left(\mu_{\ln N2} - \beta_1 \sigma_{\ln N2}\right) = 17{,}493.1. \tag{b}$$

Now, at the stress level 2, the total number of cycles n_{2-eq} including transferred equivalent number of cycles will be:

$$n_{2-eq} = n_{L2} + n_{eq1-2} = 15{,}000 + 17{,}493.1 = 32{,}493.1. \tag{c}$$

Therefore, the reliability of the specimen under two stress levels per Equation (2.49) will be:

$$R = P\left(N_2 \geq n_{2-eq}\right) = P\left(\ln\left(N_2\right) \geq \ln\left(n_{2-eq}\right)\right)$$
$$= \Phi\left(\frac{\mu_{\ln N_2} - \ln\left(n_{2-total}\right)}{\sigma_{\ln N_2}}\right) = \Phi\left(\frac{10.9551 - \ln(32{,}493.1)}{0.242377}\right)$$
$$= \Phi(2.33651) = 0.99027.$$

(2) The K-D model.

The cyclic axial stresses in this example are non-zero-mean cyclic stresses. We need to convert them into a fully reversed cyclic stress per Equation (2.21).

For the stress level #1 with $\sigma_{a1} = 20.833$ (ksi) and $\sigma_{m1} = 20.833$ (ksi), the fully reversed equivalent stress amplitude of σ_{eq1} is

$$\sigma_{eq1} = \frac{\sigma_{a1} \times S_u}{(S_u - \sigma_{m1})} = \frac{20.83333 \times 51.2}{(51.2 - 20.83333)} = 35.12623 \ (\text{ksi}). \tag{d}$$

For the stress level #2 with $\sigma_{a2} = 22.5$ (ksi) and $\sigma_{m2} = 22.5$ (ksi), the fully reversed equivalent stress amplitude of σ_{eq2} is

$$\sigma_{eq2} = \frac{\sigma_{a2} \times S_u}{(S_u - \sigma_{m2})} = \frac{22.5 \times 51.2}{(51.2 - 22.5)} = 40.139 \ (\text{ksi}). \tag{e}$$

The fatigue damage index D due to the specified cyclic axial stresses per Equation (2.85) will be:

$$D = n_{L1}\sigma_{eq1}^m + n_{L2}\sigma_{eq2}^m = 30{,}000 \times (35.126)^{3.8812} + 15{,}000 \times (40.139)^{3.8812}$$
$$= 5.50358 \times 10^{10}. \tag{f}$$

In this example, the fatigue damage index D is deterministic. The reliability of the specimen under the specified cyclic axial stress by the definition of reliability will be:

$$R = P\left(K_0 \geq D\right) = P\left(\ln\left(K_0\right) \geq \ln\left(D\right)\right) = \Phi\left(\frac{\mu_{\ln K_0} - \ln(D)}{\sigma_{\ln K_0}}\right)$$
$$= \Phi\left(\frac{25.3014 - 24.73125}{0.238106}\right) = \Phi\left(2.322869\right) = 0.98991.$$

The results from the P-S-N curve and the K-D model are almost identical. The relative difference is 0.036%. ∎

These three examples cover typical cyclic loadings. Results from these three examples have shown that the reliability obtained from the P-S-N curves and the K-D models for the same issue are almost the same with a maximum relative error 0.036%. These have approved and verified that the probabilistic fatigue damage model (the K-D model) can be used to describe the fatigue test data of the same material specimen under different fatigue test stress levels.

The K-D model can be used to solve the reliability of a component under any type of cyclic loadings. The following Example 2.35 could not be directly solved by the P-S-N curve approach and but can be solved by the K-D model.

Example 2.35

The aluminum 6061-T6 10 Gauge sheet-type fatigue specimen as shown in Figure 2.8 is subjected to cyclic axial stress with a stress amplitude $\sigma_a = 25$ (ksi) and a mean stress $\sigma_m = 15$ (ksi). The number of cycles of this cyclic axial stress is 50,000 (cycles). The ultimate tensile strength of this material is $S_u = 51.2$ (ksi). The P-S-N curves of this material is listed in Table 2.13. The K-D model is listed in Table 2.35. Use the P-S-N curve and the K-D probabilistic model to calculate its reliability.

Solution:

(1) The P-S-N curve approach.

The P-S-N curves of this material listed in Table 2.13 do not have a probabilistic distribution function for this specific cyclic axial stress which is a stress amplitude $\sigma_a = 25$ (ksi) and a mean stress $\sigma_m = 15$ (ksi). So, the P-S-N curve approach could not be directly used to calculate the reliability of this problem.

(2) The K-D model.

The K-D probabilistic fatigue damage model can be used to solve this problem.

The equivalent stress amplitude of the cyclic stress with a stress amplitude $\sigma_a = 25$ (ksi) and a mean stress $\sigma_m = 15$ (ksi) per Equation (2.21) is:

$$\sigma_{eq} = \frac{\sigma_a \times S_u}{(S_u - \sigma_m)} = \frac{25 \times 51.2}{(51.2 - 15)} = 35.359 \text{(ksi)}. \tag{a}$$

The fatigue damage index D due to the specified cyclic axial stress per Equation (2.85) will be:

$$D = n_L \sigma_{eq}^m = 50,000 \times (35.35912)^{3.8812} = 5.11698 \times 10^{10}. \tag{b}$$

In this example, the fatigue damage index D is deterministic. The reliability of the specimen under the specified cyclic axial stress will be:

$$R = P\,(K_0 \geq D) = P\,(\ln(K_0) \geq \ln(D)) = \Phi\left(\frac{\mu_{\ln K_0} - \ln(D)}{\sigma_{\ln K_0}}\right)$$

$$= \Phi\left(\frac{25.3014 - 24.65841}{0.238106}\right) = \Phi\,(2.61961) = 0.99560.$$

■

Both the P-S-N curves and the K-D model are probabilistic fatigue theory and can be used to calculate the reliability of a component under cyclic loading spectrums. There are four distinguished features of the K-D model with comparison to the P-S-N curve approach.

1. The K-D model uses all fatigue test data under all test stress levels to conduct statistical analysis to represent the scatters of material fatigue behavior. From the general view of statistics, the derived distribution parameters in the K-D model are much reliable because of the much larger sample size.

2. If sometimes, the fatigue test data of some materials are not enough to compile a practical P-S-N curve, the data may still be used by the K-D model to describe the fatigue behaviors and to conduct the calculation of the fatigue reliability. For example, five test data on each of seven different constant stress levels are not enough for the P-S-N curve approach. However, for the K-D model, the sample size is 35. Therefore, the K-D model is more practical for fatigue reliability design.

3. In the K-D model, one probabilistic distribution function is used to describe material fatigue behavior and to conduct fatigue reliability evaluation. So, it is more convenient for fatigue reliability evaluation.

4. The K-D model can be used to calculate the reliability of a component under any type of cyclic loading spectrum.

2.10   SUMMARY

Fatigue failure is one of the most common and important failure modes when a metal component is subjected to a cyclic stress spectrum. Three parameters for describing cyclic stress are stress mean σ_m, stress amplitude σ_a and the number of cycles n_L. Since the mean stress will be mainly used to calculate the equivalent fully revered stress amplitude, stress amplitude σ_a and the number of cycles n_L are two main parameters for describing a cyclic loading spectrum. Six different cyclic loading spectrums have been discussed in Section 1.2 and relisted here in Table 2.61. Any cyclic stress spectrum or cyclic loading spectrum can be described by one of these six cyclic loading spectrums.

Table 2.61: Six cyclic loading spectrums

Cyclic loading spectrum model #	Number of cycles n_L	Mean stress σ_m	Stress amplitude σ_a
#1	n_L-constant	σ_m-constant	σ_a-constant
#2	n_L-distributed	σ_m-constant	σ_a-constant
#3	n_L-constant	σ_m-constant	σ_a-distributed
#4	$n_{Li}, i = 1,2,\ldots, L$ constants	$\sigma_{mi}, i = 1,2,\ldots, L$ constants	$\sigma_{ai}, i = 1,2,\ldots, L$ constants
#5	$n_{Li}, i = 1,2,\ldots, L$ distributed	$\sigma_{mi}, i = 1,2,\ldots, L$ constants	$\sigma_{ai}, i = 1,2,\ldots, L$ constants
#6	$n_{Li}, i = 1,2,\ldots, L$ constants	$\sigma_{mi}, i = 1,2,\ldots, L$ constants	$\sigma_{ai}, i = 1,2,\ldots, L$ distributed

For a high-cycle fatigue issue, which is a common case in general engineering design, material fatigue strength data is typically obtained through a stress-life method. In a stress-life method, fatigue specimen is subjected to constant cyclic stress. The fatigue test result per a fatigue test is the number of cycles at failure, that is, a fatigue life under specified constant cyclic stress level. Both the P-S-N curve approach discussed in Section 2.8 and the K-D probabilistic fatigue damage model discussed in Section 2.9 are fatigue theories to analyze and to represent the fatigue test data for the calculation of reliability of component under a cyclic loading spectrum.

The P-S-N curve approach can provide two sets of distributions. One is the P-N distributions, in which material fatigue life under a specified constant cyclic stress level is treated as a random variable and described by a normal distribution or a log-normal distribution. Another is the P-S distributions, in which material fatigue strength at a specified constant fatigue life, is treated as a random variable and typically described by a normal distribution or a log-normal distribution. When the P-S-N curve approach is used to calculate the reliability of a component under a cyclic loading spectrum, we have the following conclusions.

- When a component is subjected to model #1, model #2, or model #3 cyclic loading spectrum, we can use the P-S-N curve to build the limit state function. Therefore, the reliability of component under such cyclic loading spectrum can be calculated by the definition of reliability, or the H-L, R-F, or Monte Carlo method. These are discussed in Sections 2.8.3, 2.8.4, and 2.8.5.

- When a component is subjected to model #4, or model #5 cyclic loading spectrum, the limit state function of a component under such cyclic loading cannot be established. However, equivalent fatigue damage concepts can be used to calculate the reliability of the component. These are discussed in Sections 2.8.6 and 2.8.7.

- When a component is subjected to model #6 cyclic loading spectrum, a series reliability block diagram is used to calculate the reliability of the component. This is discussed in Section 2.8.8.

The K-D probabilistic fatigue damage model uses all fatigue test data of the same material fatigue specimen under different cyclic stress levels to provide a three-parameters distribution model for describing material fatigue strength. These three-parameters are m—the slope of the traditional S-N curves in both log-axis scales, $\mu_{\ln k_0}$—the log-mean of and $\sigma_{\ln k_0}$—the log-standard deviation of lognormally distributed material fatigue strength index K_0. When the K-D model is used to deal with fatigue design of a component under a cyclic loading spectrum, we can establish a limit state function of a component under any of the six models of cyclic loading spectrums. Therefore, the reliability of component under such cyclic loading spectrum can be calculated by the definition of reliability, or the H-L method, R-F, or Monte Carlo method. These are discussed in Sections 2.9.5–2.9.11.

2.11   REFERENCES

[1] Lalanne, C., *Fatigue Damage*, 3rd ed., Wiley, Hoboken, NJ, 2014. DOI: 10.1002/9781118931189. 9, 72

[2] Budynas, R. G. and Nisbett, J. K., *Shigley's Mechanical Engineering Design*, 10th ed., Mc-Graw Hill Education, New York, 2014. 9, 16, 19, 27, 96

[3] Callister, W. D. Jr. and Rethwisch, D. R., *Materials Science and Engineering: An Introduction*, 9th ed., Joseph Wiley, Hoboken, NJ, 2014. 10, 72

[4] Le, Xiaobin, The reliability calculation of components under any cyclic fatigue loading spectrum, *ASME International Mechanical Engineering Congress and Exposition, IMECE–70084*, Tampa, FL, November 3–9, 2017. DOI: 10.1115/imece2017-70084. 11

[5] Ugural, A. C., *Mechanical Design of Machine Components*, 2nd ed., CRC Press, Taylor & Francis Group, Boca Raton, FL, 2015. DOI: 10.1201/9781315369679. 14, 19, 72

[6] Ling, J. and Pan, J., Engineering method for reliability analyses of mechanical structures for long fatigue lives, *Reliability Engineering and System Safety*, 56, pp. 135–142, 1997. DOI: 10.1016/s0951-8320(97)00012-4. 16

[7] Rao, S. S., *Reliability Engineering*, Person, 2015. 16, 17, 20, 30, 36, 64

[8] Le, Xiaobin, *Reliability-Based Mechanical Design, Volume 1: Component under Static Load*, Morgan & Claypool Publishers, San Rafael, CA, 2020. 22, 36, 60, 75, 77

[9] Haugen, E. B., *Probabilistic Mechanical Design*, John Wiley & Sons, Inc., 1980. 30, 36

[10] Kececioglu, D. B., Smith, R. E., and Felsted, E. A., Distributions of cycles-to-failure in simple fatigue and the associated reliabilities, *Annals of Assurance Science, 8th Reliability and Maintainability Conference*, pp. 357–374, Denver, CO, July 7–9, 1969. 30, 31, 64

[11] Kececioglu, D. B., *Robust Engineering Design-by-Reliability with Emphasis on Mechanical Components and Structural Reliability*, DEStech Publications, Inc., Lancaster, PA, 2003. 30, 31, 34, 52

[12] Le, Xiaobin, A probabilistic fatigue damage model for describing the entire set of fatigue test data of the same material, *ASME International Mechanical Engineering Congress and Exposition, IMECE–10224*, Salt Lake City, UT, November 8–14, 2019. 31

[13] ASTM STM E466—15, Standard Practice for Conducting Force Controlled Constant Amplitude Axial Fatigue Tests of Metallic Materials. DOI: 10.1201/9781420035636. 32

[14] Martinez, W. L. and Martinez, A. R., *Computational Statistics Handbook with MATLAB*, 3rd ed., CRC Press, Boca Raton, FL, 2016. DOI: 10.1201/9781420035636. 36

[15] Kececioglu, D., Probabilistic design methods for reliability and their data and research requirements, *Failure Prevention and Reliability*, pp. 285–305, ASME, 1977. 52

[16] Le, Xiaobin, The reliability of a component under multiple cyclic stress levels with distributed cyclic numbers, *ASME International Mechanical Engineering Congress and Exposition, IMECE–65269*, Phoenix, AZ, November 11–17, 2016. DOI: 10.1115/imece2016-65269. 55

[17] Le, Xiaobin, The reliability calculation of components under any cyclic fatigue loading spectrum, *ASME International Mechanical Engineering Congress and Exposition, IMECE–70084*, Tampa, FL, November 3–9, 2017. DOI: 10.1115/imece2017-70084. 64

[18] Miner, M. A., Cumulative damage in fatigue, *Journal of Applied Mechanics*, vol. 12, no. 3, pp. 159–164, 1945. 64

[19] Le, Xiaobin, Applications of the Monte Carlo method for estimating the reliability of components under multiple cyclic fatigue loadings, *ASME International Mechanical Engineering Congress and Exposition, IMECE–86130*, Pittsburg, PA, November 9–15, 2018. DOI: 10.1115/imece2018-86130. 70

[20] Suresh, S., *Fatigue of Materials*, Cambridge University Press, New York, 1998. DOI: 10.1017/cbo9780511806575. 72

[21] Henry, S. D. and Redenbatch, F., *Fatigue Data Book: Light Structure Alloys*, ASM International, Materials Park, OH, 1995. 73

[22] Le, Xiaobin, A probabilistic fatigue damage model for describing the entire set of fatigue test data of the same material, *ASME International Mechanical Engineering Congress and Exposition, IMECE–10224*, Salt Lake City, UT, November 8–14, 2019. 74, 75

[23] Zong, W. H. and Le, Xiaobin, *Probabilistic Design Method of Mechanical Components*, Shanghai Jiao Tong University Publisher, September 1995, Shanghai, China.

[24] Le, Xiaobin and Peterson, M. L., A method for fatigue-based reliability when the loading of the component is unknown, *International Journal of Fatigue*, vol. 21(6), pp. 603–610, 1999. DOI: 10.1016/S0142-1123(99)00016-X. 74, 75

2.12 EXERCISES

2.1. What is fatigue? Describe one fatigue failure example.

2.2. Explain the fatigue failure mechanism. For a component with the same type of material and dimensions under the same type of cyclic loading, which one will be failure first if one is machined with a lathe and another one is machined with a polished surface finish? Explain the choice.

2.3. A group of fatigue test for a steel specimen is listed in Table 2.62. Calculate the slope of the traditional S-N curve in both ln-axis scales.

Table 2.62: Steel specimen

Stress amplitude (ksi)	Sample size	Fatigue life (cycles) $\times 10^3$
25.14	4	328.12, 315.87, 337.62
30.38	2	64.18, 71.62
35.81	5	16.53, 16.95, 17.39, 18.03, 17.27
42.85	3	3.32, 3.85, 3.62

2.4. A group of fatigue test for a metal specimen is listed in Table 2.63. Calculate the slope of the traditional S-N curve in both ln-axis scales.

2.5. A machined bar with a diameter 2.150″ is subjected to cyclic axial loading. Its ultimate material strength is 61.5 ksi. If the fatigue strength is obtained from a standard polished specimen under a fully reversed bending stress, determine k_a, k_b, and k_c for this component.

2.6. A forged shaft with a diameter 1.750″ is subjected to cyclic bending stress. Its ultimate material strength is 91.7 ksi. If the fatigue strength is obtained from a standard polished specimen under a fully reversed bending stress, determine k_a, k_b, and k_c for this component.

Table 2.63: Metal specimen

Stress amplitude (Mpa)	Sample size	Fatigue life (cycles) $\times 10^3$
700	5	3.28, 3.81, 3.74, 3.69, 3.78
650	3	17.28, 18.09, 16.89
500	3	52.11, 49.83, 53.72
200	4	243.32, 239.97, 245.74, 244.29

2.7. A component is subjected to cyclic bending stress with a stress mean $\sigma_m = 18.38$ (ksi) and a stress amplitude $\sigma_a = 12.52$ (ksi). Its ultimate material strength is 71.8 ksi. Calculate the equivalent stress amplitude of a fully reversed cyclic stress.

2.8. A component is subjected to cyclic axial stress with a maximum stress $\sigma_{max} = 11.78$ (ksi) and a minimum stress $\sigma_{min} = -50.14$ (ksi). Its ultimate material strength is 61.5 ksi. Calculate the equivalent stress amplitude of a fully reversed cyclic stress.

2.9. A shaft shoulder with a fillet radius $r = 0.032''$ is subjected cyclic bending stress. Its theoretical stress concentration factor due to static bending stress is 1.69. The shaft material ultimate strength is 61.5 ksi. Determine its fatigue stress concentration factor.

2.10. A plate with a center transverse hole is subjected to cyclic axial loading. The radius of the hole is $0.25''$. Its theoretical stress concentration factor due to static bending stress is 2.58. The shaft material ultimate strength is 61.5 ksi. Determine its fatigue stress concentration factor.

2.11. A machined constant circular beam with a diameter $d = 1.250 \pm 0.005''$ is subjected a cyclic bending loading. The mean bending loading M_m of the cyclic bending moment is 1.72 (klb.in). The bending moment amplitude M_a of the cyclic bending moment follows a normal distribution with a mean $\mu_{M_a} = 2.26$ (klb.in) and a standard deviation $\sigma_{M_a} = 0.93$ (klb.in). The ultimate material strength is 61.5 (ksi). The component endurance limit S_e follows a normal distribution with a mean $\mu_{S_e} = 22.4$ (ksi) and a standard deviation $\sigma_{S_e} = 1.44$ (ksi). This bar is designed to have an infinite life. (1) Establish the limit state function of this beam. (2) Calculate the reliability of the beam under the cyclic bending loading.

2.12. A rectangular bar with a height $h = 1.250 \pm 0.005''$ and a width $b = 2.250 \pm 0.005''$ is subjected to cyclic axial loading with a mean $F_m = 32.15$ (klb). The axial loading amplitude F_a follows a uniform distribution between 7.15 (klb) and 9.25 (klb). The ultimate material strength is 61.5 (ksi). The component endurance limit S_e follows a normal distribution with a mean $\mu_{S_e} = 17.4$ (ksi) and a standard deviation $\sigma_{S_e} = 2.84$ (ksi). This bar is designed to have an infinite life. (1) Establish the limit state function of this problem. (2) Calculate the reliability of the bar under the cyclic bending loading.

2.13. A shaft with a diameter $d = 1.500 \pm 0.005''$ is subjected to a cyclic torsion loading. The torque can be treated as a fully reversed cyclic torsion. The torque amplitude in the unit of klb-in can be treated as lognormal distribution with a log mean $\mu_{\ln T} = 1.85$ and log standard deviation $\sigma_{\ln T} = 0.062$. The component torsion endurance limit S_e follows a normal distribution with a mean $\mu_{S_e} = 12.4$ (ksi) and a standard deviation $\sigma_{S_e} = 1.02$ (ksi). This shaft is designed to have an infinite life. (1) Establish the limit state function of this problem. (2) Calculate the reliability of the shaft under the cyclic bending loading

2.14. What is the P-S-N curve approach? What are the two sets of curves that the P-S-N curve can provide?

2.15. Conduct literature research and find an example where the P-S-N curves are presented.

2.16. A bar is subjected to cyclic axial stress with a mean stress $\sigma_m = 12.6$ (ksi) and stress amplitude $\sigma_a = 8.6$ (ksi). According to the design specification, the bar has a design life $n_L = 380,000$ (cycles). The bar fatigue life N_C at the stress level with a mean stress $\sigma_m = 12.6$ (ksi) and stress amplitude $\sigma_a = 8.6$ (ksi) follow a normal distribution with a mean $\mu_{N_C} = 440,000$ (cycles) and a standard deviation $\sigma_{N_C} = 34,500$. Calculate its reliability.

2.17. A beam is subjected to a fully reversed cyclic bending stress with a constant number of cycles $n_L = 450,000$ (cycles). The stress amplitude of this fully reversed cyclic bending stress σ_a follows a Weibull distribution with a scale parameter $\eta = 18.25$ (ksi) and a shape parameter $\beta = 1.5$. The beam fatigue strength S_f at the fatigue life $N = 450,000$ (cycles) follows a normal distribution with a mean $\mu_{S_f} = 21.98$ (ksi) and a standard deviation $\sigma_{S_f} = 1.78$ (ksi). Calculate its reliability

2.18. A square bar is subjected to cyclic fully reversed axial stress with a constant stress amplitude $\sigma_a = 24.6$ (ksi). Its number of cycles of this fully reversed axial stress can be treated as a normal distribution with a mean $\mu_{n_L} = 125,000$ (cycles) and a standard deviation $\sigma_{n_L} = 5600$ (cycles). The bar fatigue life N_C at the fatigue strength level $S_f = 24.6$ (ksi) follows a lognormal distribution with a log mean $\mu_{\ln N_C} = 12.13$ and a standard deviation $\sigma_{\ln N_C} = 0.249$. Calculate its reliability.

2.19. A round bar is subjected to three constant cyclic stresses as listed in the 2nd and 3rd columns of Table 2.64. The distributed fatigue life of this bar at the corresponding stress levels are listed in the 4th and 5th columns of Table 2.64. Calculate its reliability.

2.20. A beam is subjected to three constant cyclic bending stresses with a distributed number of cycles as listed in the 2nd, 3rd, and 4th columns of Table 2.65. The distributed fatigue life of this bar at the corresponding stress levels are listed in the 5th and 6th columns of Table 2.65. Calculate its reliability.

Table 2.64: Cyclic stresses

Cyclic Loading Stress Spectrum			Component P-N Distributions	
Stress Level #	σ_{ai} (ksi)	n_{Li}	Fatigue Life N_{Ci} (normal distribution)	
			$\mu_{N_{Ci}}$	$\sigma_{N_{Ci}}$
1	25	90,800	151,000	12,100
2	35	4,500	8,800	790
3	40	1,825	2,830	235

Table 2.65: Cyclic bending stresses

Cyclic Loading Stress Spectrum				Component P-N Distributions	
Stress Level #	σ_{ai} (ksi)	Number of Cycles n_{Li} (normal distribution)		Fatigue Life N_{Ci} (normal distribution)	
		$\mu_{n_{Li}}$	$\sigma_{n_{Li}}$	$\mu_{N_{Ci}}$	$\sigma_{N_{Ci}}$
1	25	80,900	5,900	151,000	12,100
2	35	4,050	398	8,800	790
3	40	1,625	150	2,830	235

2.21. A shaft is subjected to model #6 cyclic bending stress. The numbers of cycles of each stress level are constant and are listed in the 2nd column of Table 2.66. The fully reversed bending stress follows a normal distribution, and their distribution parameters are listed in the 3rd and 4th columns of Table 2.66. The component fatigue strengths S_f at the corresponding fatigue life are normally distributed random variable, and their distribution parameters are listed in the 5th and 6th columns of Table 2.66. Calculate its reliability.

Table 2.66: Cyclic bending stresses

Cyclic Loading Stress Spectrum				Component P-S_{fi} Distributions	
Stress Level #	n_{Li} (ksi)	Stress Amplitude σ_{ai} (normal distribution)		Fatigue Strength S_{fi} (normal distribution)	
		$\mu_{\sigma_{ai}}$	$\sigma_{\sigma_{ai}}$	$\mu_{S_{fi}}$	$\sigma_{S_{fi}}$
1	3,000	27.15	2.12	39.77	3.18
2	20,000	21.5	2.19	31.78	2.86
3	300,000	18.34	1.25	23.06	1.73

2.22. Conduct literature research to find a fatigue test under at least three different stress level and with more than 30 tests. Then use the K-D probabilistic model to determine its three distribution parameters: m, $\mu_{\ln K_0}$, and $\sigma_{\ln K_0}$.

2.23. A bar with a diameter $1.125 \pm 0.005''$ is subjected to a constant cyclic axial loading, which is listed in the first three columns of Table 2.67. The ultimate material strength S_u is 75 (ksi). The component fatigue strength index K on the critical section for the axial cyclic loading is listed in the last three columns of Table 2.67. For the component fatigue strength index K, the stress unit is ksi. Calculate the reliability of this bar.

Table 2.67: Cyclic axial loading

Cyclic Axial Loading			Component Fatigue Strength Index K (lognormal distribution)		
n_L	F_m	F_a	m	$\mu_{\ln K}$	$\sigma_{\ln K}$
480,000	20.74 (klb)	18.82 (klb)	8.21	41.738	0.357

2.24. A bar with a diameter $1.500 \pm 0.005''$ is subjected to a fully reversed cyclic axial loading. The axial loading amplitude follows a normal distribution. Information of this cyclic axial loading is listed in the first three columns of Table 2.68. The component fatigue strength index K on the critical section for the axial cyclic loading is listed in the last three columns of Table 2.68. For the component fatigue strength index K, the stress unit is ksi. Calculate the reliability of this bar.

Table 2.68: Cyclic axial loading

Cyclic Axial Loading			Component Fatigue Strength Index K (lognormal distribution)		
n_L	Normally Distributed F_a		m	$\mu_{\ln K}$	$\sigma_{\ln K}$
	μ_{F_a}	σ_{F_a}			
250,000	50.68 (klb)	4.82 (klb)	8.21	41.738	0.357

2.25. A round bar with a diameter $1.250 \pm 0.005''$ is subjected to a fully reversed cyclic bending loading with a constant bending moment amplitude. The number of cycles of this cyclic bending loading follows a normal distribution. Information of this cyclic bending loading is listed in the first three columns of Table 2.69. The component fatigue strength index K on the critical section for the cyclic bending loading is listed in the last three columns of Table 2.69. For the component fatigue strength index K, the stress unit is ksi. Calculate the reliability of this bar.

Table 2.69: Reverse cyclic bending loading

Cyclic Bending Loading				Component Fatigue Strength Index K (lognormal distribution)		
Normally Distributed n_L		M_a	m	$\mu_{\ln K}$	$\sigma_{\ln K}$	
μ_{n_L}	σ_{n_L}					
200,000	8,900	8.86 (klb/in)	8.21	41.738	0.357	

2.26. A square beam with a side height $1.75 \pm 0.005''$ is subjected to several cyclic bending loading. The number of cycles and the bending moment amplitudes of this cyclic bending loading are all constants. Information of this fully reversed bending loading is listed in the first three columns of Table 2.70. The component fatigue strength index K on the critical section for the cyclic bending loading is listed in the last three columns of Table 2.70. For the component fatigue strength index K, the stress unit is ksi. Calculate the reliability of this beam.

Table 2.70: Several cyclic bending loading

Cyclic Bending Loading			Component Fatigue Strength Index K (lognormal distribution)		
n_L	M_m (klb.in)	M_a (klb.in)	m	$\mu_{\ln K}$	$\sigma_{\ln K}$
3,500	10.64	18.24			
30,000	10.64	14.73	8.21	41.738	0.357
450,000	10.64	8.75			

2.27. A square beam with a side height $1.500 \pm 0.005''$ is subjected to several fully reversed cyclic bending loading. The number of cycles in each loading level is constants. The bending moment amplitudes follow normal distributions. Information of this fully reversed bending loading is listed in the first three columns of Table 2.71. The component fatigue strength index K on the critical section for the cyclic bending loading is listed in the last three columns of Table 2.71. For the component fatigue strength index K, the stress unit is ksi. Calculate the reliability of this bar.

2.28. A shaft with a diameter $1.125 \pm 0.005''$ is subjected to several fully reversed cyclic torsion loading. The torsion amplitude in each loading level is constant. The number of cycles in each torsion level follows a normal distribution. Information of this fully reversed torsion loading is listed in the first three columns of Table 2.72. The component fatigue strength index K on the critical section for the cyclic torsion loading is listed in

Table 2.71: Fully reversed cyclic bending loading

Cyclic Bending Loading			Component Fatigue Strength Index K (lognormal distribution)		
n_L	Normally Distrubuted M_a		m	$\mu_{\ln K}$	$\sigma_{\ln K}$
	μ_{M_a} (klb.in)	σ_{M_a} (klb.in)			
2,500	16.51	1.49			
40,000	11.08	0.89	8.21	41.738	0.357
350,000	7.86	0.53			

Table 2.72: Fully reversed cyclic torsion loading

Cyclic Torsion Loading			Component Fatigue Strength Index K (lognormal distribution)		
Normally Distributed n_L		T_a (klb.in)	m	$\mu_{\ln K}$	$\sigma_{\ln K}$
μ_{n_L}	σ_{n_L}				
2,000	160	8.84			
560,000	35,400	3.87	8.21	37.308	0.518

the last three columns of Table 2.72. For the component fatigue strength index K, the stress unit is ksi. Calculate the reliability of this shaft.

2.29. The critical section of a solid rotating shaft with a diameter $1.250 \pm 0.005''$ is subjected to cyclic combined loading. According to the design specification, the cyclic loading spectrum can be described by several stress levels and listed in the first three columns of Table 2.73. The material ultimate tensile strength S_u of the shaft is 75 (ksi). The component fatigue strength index K on the critical section for the cyclic bending stress is listed in the last three columns of Table 2.73. For the component fatigue strength index K, the stress unit is ksi. Calculate the reliability of the shaft.

Table 2.73: Cyclic combined loading

Cyclic Axial Loading			Component Fatigue Strength Index K (lognormal distribution)		
n_L	T_i (klb.in)	M_i (klb.in)	m	$\mu_{\ln K}$	$\sigma_{\ln K}$
3,000	4.78	7.32			
200,000	4.78	4.16	8.21	41.738	0.357

2.30. The critical section of a solid rotating shaft with a diameter $2.250 \pm 0.005''$ is subjected to cyclic combined loading. According to the design specification, the cyclic loading spectrum can be described by several stress levels with distributed stress amplitudes and are listed in the first four columns of Table 2.74. The material ultimate tensile strength S_u of the shaft is 75 (ksi). The component fatigue strength index K on the critical section for the cyclic bending stress is listed in the last three columns of Table 2.74. For the component fatigue strength index K, the stress unit is ksi. Calculate the reliability of the rotating shaft.

Table 2.74: Cyclic combined loading

Cyclic Combined Loading				Component Fatigue Strength Index K (lognormal distribution)		
n_L	T (klb.in)	Normally Distributed M		m	$\mu_{\ln K}$	$\sigma_{\ln K}$
		μ_M (klb.in)	σ_M (klb.in)			
20,000	21.83	34.92	2.56	8.21	41.738	0.357
400,000	21.83	20.72	1.94			

CHAPTER 3

The Dimension of a Component with Required Reliability

3.1 INTRODUCTION

The reliability-based mechanical component design includes two main tasks: design check and dimension design. The design check is to calculate the reliability of a component under specified loading conditions, which have been discussed in Chapter 4 for static loading of the book [1] and Chapter 2 for cyclic loading spectrum of this book. Dimension design of a component is to design the component dimension with required reliability under specified loading condition. The dimension design will be the topic of this chapter.

In Section 3.2, we will first discuss how to design the component dimension with required reliability under specified loading condition. Then, in Section 3.3, we will discuss the dimension design with required reliability under static loading. Finally, in Section 3.4, we will discuss the dimension design with required reliability under cyclic loading spectrum. In each loading case, we will discuss simple loading such as axial loading, direct shearing, torsion, and bending. Then, for static loading, we will discuss component dimension design with required reliability under several typical combined loadings. However, for cyclic loading, we will only discuss the dimension design of a rotating shaft the required reliability under the combined loading of steady torsion with a constant or distributed bending moment.

3.2 DIMENSION DESIGN WITH REQUIRED RELIABILITY

3.2.1 LIMIT STATE FUNCTION AND PRELIMINARY DESIGN

Component dimension d is treated as a normally distributed random variable. Its standard deviation σ_d is determined by the manufacturing process. In mechanical component design, we typically already choose the manufacturing process before we start to calculate its dimension. So, the standard deviation σ_d of component's dimension can be treated as a known value. Therefore, component dimension design with the required reliability under specified loading is mainly to determine the mean μ_d of the dimension d. For dimension design with the required reliability

R under a specified loading condition, we will use the following general limit state function:

$$g(X_1, \ldots, X_n, d) = \begin{cases} > 0 & \text{Safe} \\ = 0 & \text{Limit state} \\ < 0 & \text{Failure,} \end{cases} \tag{3.1}$$

where X_i $(i = 1, 2, \ldots, n)$ is a random variable related to component strength or loading or component geometric dimensions, which could be any type of distributions. d is a normal distribution dimension to be designed with a mean μ_d and a standard deviation σ_d. For component dimension design, μ_d is to be solved from the limit state function (3.1). Since the component dimension with μ_d will have the required reliability R, the reliability index β from Equation (3.1) will be:

$$\beta = \Phi^{-1}(R) = norminv(R), \tag{3.2}$$

where $\Phi^{-1}(\cdot)$ is the standard normal inversed CDF. $norminv(\cdot)$ is a function in MATLAB for the standard normal inversed CDF.

For component dimension design, some design parameters are dimension-dependent. For a component dimension design under static loading, the static stress concentration factor is a dimension-dependent parameter. For a component dimension design under cyclic loading spectrum, fatigue stress concentration factor and the size modification factor are dimension-dependent parameters.

For a static stress concentration factor, we can use Tables 3.1–3.3 to estimate the preliminary stress concentration factor. In Table 3.1, r is the radius of the fillet, and d is the diameter of the smaller shaft. In Tables 3.2 and 3.3, r is the radius of the fillet, and W is the smaller width of the bar or plate. The static stress concentration factor is mainly determined by the radius of the fillet. Generally, we have the sketched structure of the component and know the reason why there is a stress concentration area. Therefore, we know the radius of the fillet even we do not know the component's dimension. When the fillet radius is determined by other mating purchased component such as a purchased bearing, we could call it a sharp radius, as shown in Tables 3.1 and 3.2. When the fillet radius is not restricted by another component, we might call a well-rounded radius, as shown in Tables 3.1 and 3.2. We can use Equation (1.3) discussed in Section 1.1 to determine the mean and the standard deviation of the static stress concentration factor.

Table 3.1: The preliminary stress concentration factors for a shaft [2]

Shaft	Bending Loading	Torsion Loading	Axial Loading
Shoulder fillet–sharp (r/d = 0.02)	2.7	2.2	3.0
Shoulder fillet–well-rounded (r/d = 0.1)	1.7	1.6	1.9

Table 3.2: The preliminary stress concentration factors for a stepped bar/plate

Stepped Plate or Bar	Bending	Axial
Shoulder fillet–sharp (r/W = 0.02)	2.7	3.0
Shoulder fillet–well-rounded (r/W = 0.1)	1.7	1.9

Table 3.3: The preliminary stress concentration factors for a plate with a center hole

Plate with a Center Hole	Bending Loading	Axial Loading
Narrow plate (d/W = 2)	1	2.1
Plate (d/W = 3)	0.667	2.3

In fatigue design, the fatigue stress concentration factor is a dimension-dependent which will be determined by dimension, fillet radius, and type of material. The fatigue stress concentration factor can be calculated per Equations (2.22), (2.23), and (2.24) in Section 2.6 when the static stress concentration factor, fillet radius, and ultimate strength are known. Tables 3.1, 3.2, and 3.3 can help to determine the preliminary static stress concentration factor. As we mentioned above, the fillet radius can be pre-determined by the type of fillets and its purpose according to the sketched structure of the component. We can also pre-select the type of material. Therefore, we can determine a preliminary estimation of the fatigue stress concentration factor.

In fatigue design, the size modification factor is also a dimension-dependent parameter and can be calculated per Equation (2.17) in Section 2.4. The size dimension factor for bending or torsion has a value between 0.807–1.12. We can use 0.87 as its preliminary estimation:

$$k_b = \begin{cases} 0.87 & \text{For bending or torsion loading} \\ 1 & \text{For axial loading.} \end{cases} \tag{3.3}$$

In the following sections, we will discuss how to determine the mean μ_d of a normally distributed dimension with the required reliability R.

3.2.2   DIMENSION DESIGN BY THE FOSM METHOD

When all random variables are normally distributed random variables in the limit state function (1.1), we can use the FOSM method to estimate μ_d with the required reliability R under specified loading condition. The FOSM method for calculating the reliability of a component has been discussed in Section 3.5 of Volume 1 [1]. Now, we will discuss how to use the FOSM to conduct the dimension design with the required reliability. The procedure of the dimension design by the FOSM method is as follows.

Step 1: Preliminary design for determining K_t for static design or K_f and k_b for fatigue design.

For a component dimension design under static loading, we can determine the preliminary static stress concentration factor, as discussed in Section 3.2.1 if the critical section is in the stress concentration area. For a component dimension design under cyclic fatigue loading spectrum, we can determine the preliminary fatigue stress concentration factor, as discussed in Section 3.2.1 if the critical section is in the stress concentration area. For the size modification factor, we will use 0.87 for bending and torsion or 1 for axial loading per Equation (3.3).

Step 2: Use the FOSM method to calculate μ_d^{*0}.

According to the FOSM method, the mean and standard deviation of the limit state function [1] will be:

$$\mu_g = g\left(\mu_{X_1}, \ldots, \mu_{X_n}, \mu_d^{*0}\right) \tag{3.4}$$

$$\sigma_g = \sqrt{\sum_{i=1}^{n}\left(\left.\frac{\partial g(X_1, \ldots, X_n, d)}{\partial X_i}\right|_{\text{means}} \times \sigma_{X_i}\right)^2 + \left(\left.\frac{\partial g(X_1, \ldots, X_n, d)}{\partial d}\right|_{\text{means}} \times \sigma_d\right)^2}, \tag{3.5}$$

where μ_{X_i} and σ_{X_i} are the mean and the standard deviation of normally distributed X_i. μ_g and σ_g are the mean and the standard deviation of the limit state function $g(X_1, \ldots, X_n, d)$. σ_d is the standard deviation of the dimension and is pre-determined by the dimension tolerance per Equation (1.1).

For dimension design, the reliability of a component is given; that is, $R = \Phi(\beta)$. Therefore, the design equation for a dimension with the required reliabilit R by using the FOSM method [1] is

$$\beta = \Phi^{-1}(R) = \frac{\mu_g}{\sigma_g}$$

$$= \frac{g\left(\mu_{X_1}, \ldots, \mu_{X_n}, \mu_d^{*0}\right)}{\sqrt{\sum_{i=1}^{n}\left(\left.\frac{\partial g(X_1, \ldots, X_n, d)}{\partial X_i}\right|_{\text{means}} \times \sigma_{X_i}\right)^2 + \left(\left.\frac{\partial g(X_1, \ldots, X_n, d)}{\partial d}\right|_{\text{means}} \times \sigma_d\right)^2}}. \tag{3.6}$$

In Equation (3.6), μ_d^{*0} is the only one unknown and can be solved.

Step 3: Update the dimension dependent parameters.

If there are dimension-dependent parameters in the limit state function, we need to use the iterative process. After a new value μ_d^{*0} of the mean μ_d in an iterative step is available, we can use μ_d^{*0} to update dimension-dependent parameters such as the static stress concentration factor K_t, the fatigue stress concentration factor K_f, or the size modification factor k_b if necessary. Then, we can go back to Step 2 to calculate a new dimension μ_d^{*1}.

Step 4: Convergence condition.

For a dimension design, since we will obtain an approximate result, we can use the following convergence condition:

$$abs(\mu_d^{*1} - \mu_d^{*0}) < 0.001''. \tag{3.7}$$

If the convergence condition is not satisfied, we go back to Step 2 until the convergence condition is satisfied.

If the convergence condition (3.7) is satisfied, the μ_d^{*1} will be the mean of the dimension with the required reliability, that is,

$$\mu_d = \mu_d^{*1}. \tag{3.8}$$

If the limit state function is a linear function of all normally distributed random variables, the FOSM method will provide an accurate result. However, if the limit state function is a nonlinear function of all normally distributed random variables, the FOSM method will only provide an approximate result [1].

Example 3.1
A circular stepped bar as shown in Figure 3.1 is subjected to axial loading F, which follows a normal distribution with a mean $\mu_F = 28.72$ (klb) and a standard deviation $\sigma_F = 2.87$ (klb). The material of this bar is ductile. The yield strength S_y of this bar's material follows a normal distribution a mean $\mu_{S_y} = 32.2$ (ksi) and a standard deviation $\sigma_{S_y} = 3.63$ (ksi). Determine the diameter d of the bar with a reliability 0.99 when its dimension tolerance is ± 0.005.

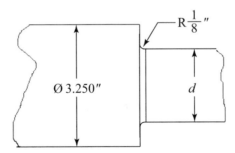

Figure 3.1: Schematic of the segment of a stepped bar.

Solution:

(1) Preliminary design for determining K_t.

This is a static design problem. According to the schematic of the stepped shaft, we can assume that it has a well-round fillet. Per Table 3.2, we have the preliminary static stress concentration factor,

$$K_t = 1.9. \tag{a}$$

(2) The limit state function of this stepped bar.

For this problem, the critical section will be on the stress concentration section, that is, the stepped section. The normal stress on the critical section of the bar caused by the axial loading F is

$$\sigma = K_t \frac{F}{\pi d^2/4} = K_t \frac{4F}{\pi d^2}. \tag{b}$$

The limit state function of the bar is

$$g_1\left(S_y, F, K_t, d\right) = S_y - K_t \frac{4F}{\pi d^2} = \begin{cases} > 0 & \text{Safe} \\ 0 & \text{Limit state} \\ < 0 & \text{Failure.} \end{cases} \tag{c}$$

We will use "allowable force" to form another version of the limit state function as shown in Equation (d), which will be much easier to solve μ_d^{*0}

$$g_2\left(S_y, F, K_t, d\right) = S_y \left(\frac{\pi d^2}{4}\right) - K_t F = \begin{cases} > 0 & \text{Safe} \\ 0 & \text{Limit state} \\ < 0 & \text{Failure.} \end{cases} \tag{d}$$

In the limit state function, there are four normally distributed variables. The mean and standard deviation of the static stress concentration factor K_t can be determined per Equation (1.3). For the dimension, the dimension standard deviation σ_d can be determined per Equation (1.1). Their distribution parameters in the limit state function (d) are listed in Table 3.4.

Table 3.4: Distribution parameters for Example 3.1

S_y (ksi)		F (klb)		K_t		d (in)	
μ_{S_y}	σ_{S_y}	μ_F	σ_F	μ_{K_t}	σ_{K_t}	μ_d	σ_d
32.2	3.63	28.72	2.87	1.9	0.095	μ_d	0.00125

(3) μ_d^{*0} in the first iterative process.

Per Equations (3.4) and (3.5), the mean and standard deviation of the limit state function are:

$$\mu_{g2} = g_2\left(\mu_{S_y}, \mu_F, \mu_{K_t}, \mu_d\right) = \mu_{S_y} \frac{\pi \left(\mu_d^{*0}\right)^2}{4} - \mu_{K_t} \mu_F \tag{e}$$

$$\sigma_{g2} = \sqrt{\left(\frac{\pi \left(\mu_d^{*0}\right)^2}{4} \sigma_{S_y}\right)^2 + \left(-\mu_{K_t} \sigma_F\right)^2 + \left(-\mu_F \sigma_{K_t}\right)^2 + \left(\mu_{S_y} \frac{\pi \mu_d^{*0}}{2} \sigma_d\right)^2}. \tag{f}$$

The reliability index β with the required reliability $R = 0.99$ per Equation (3.2) is

$$\beta = \Phi^{-1}(0.99) = 2.32635. \tag{g}$$

Per Equation (3.6), we have:

$$\beta = \frac{\mu_{g2}}{\sigma_{g2}} = \frac{\mu_{S_y}\dfrac{\pi\left(\mu_d^{*0}\right)^2}{4} - \mu_{K_t}\mu_F}{\sqrt{\left(\dfrac{\pi\left(\mu_d^{*0}\right)^2}{4}\sigma_{S_y}\right)^2 + (-\mu_{K_t}\sigma_F)^2 + (-\mu_F\sigma_{K_t})^2 + \left(\mu_{S_y}\dfrac{\pi\mu_d^{*0}}{2}\sigma_d\right)^2}}. \tag{h}$$

In Equation (h), only μ_d^{*0} is an unknown and can be solved. Equation (h) is based on the limit state function (d). This is much simpler than the corresponding equation based on the limit state function (c). Per Equation (h) with the current value, we can solve μ_d^{*0}:

$$\mu_d^{*0} = 1.7771''. \tag{i}$$

(4) Update K_t based on $\mu_d^{*0} = 1.7771''$.

After we have dimension $\mu_d^{*0} = 1.7771$, we have the geometric dimensions for the stress concentration. Then we can calculate the static stress concentration factor K_t based on current dimensions, that is, $D = 3.25''$, $d = \mu_d^{*0} = 1.7771''$, and $r = 1/8''$. The updated K_t in this case is:

$$K_t = 2.2039. \tag{j}$$

(5) Create a MATLAB program to conduct the iterative process.

We can follow the procedure discussed above and the formula listed in this problem to make a MATLAB program. The results of the iterative process from the program is listed in Table 3.5.

From the iterative results, the dimension of d in this example is

$$d = 1.919 \pm 0.005''.$$

■

Example 3.2
On the critical cross-section of a beam with a rectangular cross-section is subjected to a bending moment $M = 50.25 \pm 4.16$ (klb.in). The yield strength S_y of this beam's material follows a normal distribution a mean $\mu_{S_y} = 32.2$ (ksi) and a standard deviation $\sigma_{S_y} = 3.63$ (ksi). If the width of the beam is $b = 2.000 \pm 0.010''$. Determine the height h of the beam with a reliability 0.95 when the dimension tolerance is $\pm 0.010''$.

Table 3.5: The results of the interactive process for Example 3.1

Iterative #	K_t	μ_d^*	$\Delta\|\mu_d^*\|$
1	1.9	1.777084	
2	2.203939	1.913948	0.136865
3	2.214874	1.918691	0.004742
4	2.21526	1.918858	0.000167

Solution:

For this example, we do not have a static stress concentration factor. So, we can directly use Step 2 of the procedure discussed above to conduct the component dimension design with the required reliability.

(1) The limit state function.

The normal stress of the beam caused by the bending moment M is

$$\sigma = \frac{Mh/2}{bh^3/12} = \frac{6M}{bh^2}. \tag{a}$$

The limit state function of the beam is

$$g(S_y, M, b, h) = S_y - \frac{6M}{bh^2} = \begin{cases} > 0 & \text{Safe} \\ 0 & \text{Limit state} \\ < 0 & \text{Failure}. \end{cases} \tag{b}$$

The bending moment M can be treated as a normal distribution. Its mean and standard deviation can be determined per Equation (1.2). For the dimension, the dimension standard deviation σ_d can be determined per Equation (1.1). In the limit state function, there are four normally distributed variables. Their distribution parameters in the limit state function (b) are listed in Table 3.6.

Table 3.6: Distribution parameters for Example 3.2

S_y (ksi)		M (klb.in)		b (in)		h (in)	
μ_{S_y}	σ_{S_y}	μ_M	σ_M	μ_b	σ_b	μ_h	σ_h
32.2	3.63	50.25	1.04	2.000	0.0025	μ_h	0.0025

(2) The mean and the standard deviation of the limit state function.

Per Equations (3.3) and (3.4), the mean and the standard deviation of the limit state function are:

$$\mu_g = g\left(\mu_{S_y}, \mu_M, \mu_b, \mu_h\right) = \mu_{S_y} - \frac{6\mu_M}{\mu_b\mu_h^2} = 32.2 - \frac{6 \times 50.25}{2.000\mu_h^2} = 32.2 - \frac{150.75}{\mu_h^2} \tag{c}$$

$$\sigma_g = \sqrt{\left(\sigma_{S_y}\right)^2 + \left(-\frac{6\sigma_M}{\mu_b\mu_h^2}\right)^2 + \left(\frac{6\mu_M\sigma_b}{\mu_b^2\mu_h^2}\right)^2 + \left(\frac{12\mu_M\sigma_h}{\mu_b\mu_h^3}\right)^2}$$

$$= \sqrt{13.1769 + \frac{9.7344}{\mu_h^4} + \frac{0.03551}{\mu_h^4} + \frac{0.56814}{\mu_h^6}} = \sqrt{13.1769 + \frac{9.7699}{\mu_h^4} + \frac{0.56814}{\mu_h^6}}. \tag{d}$$

(3) The diameter of the beam with a reliability 0.95.

The reliability index β with the required reliability $R = 0.95$ per Equation (3.2) is

$$\beta = \Phi^{-1}(0.95) = 1.64485. \tag{e}$$

Per Equation (3.5), we have the following equation:

$$1.64485 = \frac{32.2 - \dfrac{29.125}{\mu_h^2}}{\sqrt{113.1769 + \dfrac{9.7699}{\mu_h^4} + \dfrac{0.56814}{\mu_h^6}}} = \frac{32.2\mu_h^3 - 150.75\mu_h}{\sqrt{13.1769\mu_h^6 + 9.7699\mu_h^2 + 0.56814}}. \tag{f}$$

By solving Equation (f), we have:
$$\mu_h = 2.400''. \tag{g}$$

Therefore, the height of the beam with the required reliability 0.95 under the specified loading will be
$$h = 2.400 \pm 0.010''.$$

∎

3.2.3   DIMENSION DESIGN BY THE MODIFIED H-L METHOD

When a limit state function of a component is a nonlinear function of all normal distributions, we can use the modified H-L method to determine the component dimension with the required reliability under specified loading [3–5]. The H-L method iteratively calculates the reliability index β and then uses converged reliability index β to calculate the reliability, which has been discussed in Section 3.6 of Volume 1 [1]. It is also displayed in Appendix A.1 of this book. For a dimension design, we already know the reliability index β, and we will use the following modified H-L method to determine iteratively the mean μ_d of the dimension.

The general procedure for the modified H-L method for dimension design is explained and displayed as follows.

Step 1: Preliminary design for determining K_t for static or K_f and k_b for fatigue design.

Per Section 3.2.1, we can determine the dimension-dependent parameters K_t for static or K_f and k_b for fatigue issue if necessary.

Step 2: Establish the limit state function.

The general limit state function of a component for dimension design is listed in Equation (3.1) and redisplayed here:

$$g(X_1, \ldots, X_n, d) = \begin{cases} > 0 & \text{Safe} \\ = 0 & \text{Limit state} \\ < 0 & \text{Failure,} \end{cases} \tag{3.1}$$

where X_i ($i = 1, 2, \ldots, n$) is a random variable related to component strength or loading, which could be any type of distributions. d is a normal distribution dimension with a mean μ_d and a standard deviation σ_d. For component dimension design, μ_d is the variable for solving and σ_d is determined by the manufacturing process and is treated as a known variable.

Step 3: Calculate the reliability index β.

According to the required reliability of the component, we can determine the reliability index β per Equation (3.2) and is redisplayed here:

$$\beta = \Phi^{-1}(R) = norminv(R). \tag{3.2}$$

Step 4: Pick an initial design point $P^{*0}\left(X_1^{*0}, \ldots, X_n^{*0}, \ldots, d^{*0}\right)$.

The initial design point must be on the surface of the limit state function as specified by Equation (3.1). We can use the mean values for the first n variables and then use the limit state function (3.1) to determine d^{*0}:

$$\begin{aligned} X_i^{*0} &= \mu_{X_i} \qquad i = 1, 2, \ldots, n \\ g\left(X_1^{*0}, \ldots, X_n^{*0}, d^{*0}\right) &= 0. \end{aligned} \tag{3.9}$$

When the actual limit state function is provided, we can rearrange the second equation in Equation (3.9) and express d^{*0} by using $X_1^{*0}, X_2^{*0}, \ldots$ and X_n^{*0}. Let's use the following equation to represent this:

$$d^{*0} = g_1\left(X_1^{*0}, X_2^{*0}, \ldots, X_n^{*0}\right). \tag{3.10}$$

Step 5: Calculate the new design point $P^{*1}\left(X_1^{*1}, \ldots, X_n^{*1}, d^{*1}\right)$.

Based on the H-L method that was discussed in Section 3.6 of Volume 1 [1] and also concisely displayed in Appendix A.1 of this book, we can calculate the Taylor Series coefficients:

$$G_i|_{P*0} = \sigma_{X_i} \left. \frac{\partial g(X_1, \ldots, X_n, d)}{\partial X_i} \right|_{P*0} \qquad i = 1, 2, \ldots, n$$

$$G_d|_{P*0} = \sigma_d \left. \frac{\partial g(X_1, \ldots, X_n, d)}{\partial d} \right|_{P*0}. \tag{3.11}$$

Let us use variable G_0 to represent the following equation:

$$G_0 = \sqrt{\sum_{i=1}^{n} (G_i|_{P*0})^2 + (G_d|_{P*0})^2}. \tag{3.12}$$

Since the reliability index β is a known value, we can use the same approach shown in the H-L method to calculate the new design point $P^{*1}\left(X_1^{*1}, \ldots, X_n^{*1}, d^{*1}\right)$:

$$X_i^{*1} = \mu_{X_i} + \sigma_{X_i} \times \beta \times \left(\frac{-G_i|_{P*0}}{G_0} \right) \qquad i = 1, \ldots, n. \tag{3.13}$$

d^{*1} will be determined by the limit state function per Equation (3.10) and can be displayed as follows:

$$d^{*1} = g_1\left(X_1^{*1}, X_2^{*1}, \ldots, X_n^{*1}\right). \tag{3.14}$$

Step 6: Update the dimension-dependent parameters.

If there are dimension-dependent parameters such as static stress concentration factor K_t for static loading, the fatigue stress concentration factor K_f and the size factor k_b for cyclic loadings, we need to use the new dimensions to update these dimension-dependent parameters. Since the value of any random variable at the design point in the H-L method is determined by this equation:

$$d^{*1} = \mu_d + \sigma_d \times \beta \times \left(\frac{-G_d|_{P*0}}{G_0} \right).$$

Rearranging the above equation, we can get the equation for μ_d:

$$\mu_d = d^{*1} - \sigma_d \times \beta \times \left(\frac{-G_d|_{P*0}}{G_0} \right), \tag{3.15}$$

where $\left(\frac{-G_d|_{P*0}}{G_0} \right)$ will be the value calculated in Equations (3.11) and (3.12).

After this μ_d is known, the geometric dimensions for the stress concentration areas are all known. We can update the static stress concentration factor K_t. Then, we can use Equations (2.22), (2.23), and (2.24) to update the fatigue stress concentration factor K_f and use Equation (2.17) to update the size modification factor k_b.

Step 7: Check convergence condition and the mean μ_d of the dimension d.

d is the dimension in the unit of inch. Therefore, the convergence condition for the dimension d can be:

$$abs\left(d^{*1} - d^{*0}\right) < 0.0001''. \tag{3.16}$$

If the convergence condition (3.16) is not satisfied, we need to update the design point by using the following recurrence of Equation (3.17) and go back to Step 5:

$$\begin{aligned} X_i^{*0} &= X_i^{*1} \qquad i = 1, \ldots, n \\ d^{*0} &= d^{*1}. \end{aligned} \tag{3.17}$$

If the convergence condition (3.16) is satisfied, the μ_d in Equation (3.15) is the mean μ_d of the dimension d with the required reliability under the specified loading.

Since the modified H-L method is an iterative process, we should use the program for calculation. The program flowchart of the modified H-L method is shown in Figure 3.2.

Example 3.3

Use the modified H-L method to do Example 3.1.

A circular stepped bar as shown in Figure 3.1 is subjected to axial loading F, which follows a normal distribution with a mean $\mu_F = 28.72$ (klb) and a standard deviation $\sigma_F = 2.87$ (klb). The material of this bar is ductile. The yield strength S_y of this bar's material follows a normal distribution a mean $\mu_{S_y} = 32.2$ (ksi) and a standard deviation $\sigma_{S_y} = 3.63$ (ksi). Determine the diameter d of the bar with a reliability 0.99 when its dimension tolerance is ± 0.005.

Solution:

(1) Preliminary design for determining K_t.

This is a static design problem. According to the schematic of the stepped shaft, we can assume that it has a well-rounded fillet. Per Table 3.2, we have the preliminary static stress concentration factor

$$K_t = 1.9. \tag{a}$$

(2) The limit state function.

The normal stress of the bar caused by the axial loading F is

$$\sigma = K_t \frac{F}{\pi d^2 / 4} = K_t \frac{4F}{\pi d^2}. \tag{b}$$

The limit state function of the bar is

$$g\left(S_y, F, K_t, d\right) = S_y - K_t \frac{4F}{\pi d^2} = \begin{cases} > 0 & \text{Safe} \\ 0 & \text{Limit state} \\ < 0 & \text{Failure.} \end{cases} \tag{c}$$

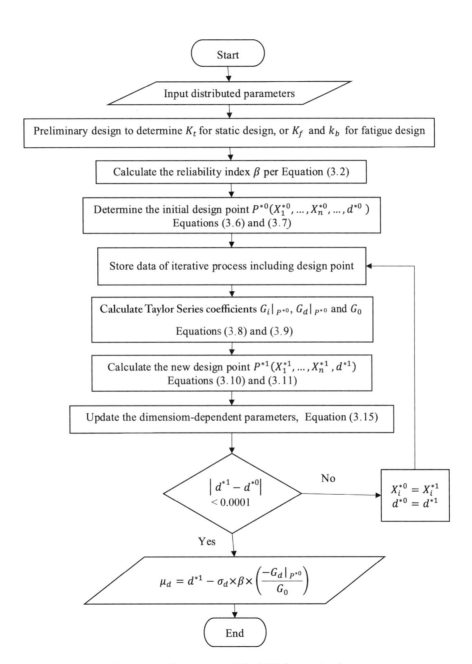

Figure 3.2: The program flowchart for the modified H-L method.

In the limit state function, there are four normally distributed variables. For the dimension, the dimension standard deviation σ_d can be determined per Equation (1.1). The mean and standard deviation of the static stress concentration factor K_t are determined per Equation (1.3). Their distribution parameters in the limit state function (c) are listed in Table 3.7. The mean and standard deviation of K_t in the table will be updated in each iterative step. (Note: The user-defined subroutine for updating the stress concentration factor is required for this example.)

Table 3.7: Distribution parameters for Example 3.3

S_y (ksi)		F (klb)		K_t		d (in)	
μ_{S_y}	σ_{S_y}	μ_F	σ_F	μ_{K_t}	σ_{K_t}	μ_d	σ_d
32.2	3.63	28.72	2.87	1.9	0.095	μ_d	0.00125

(3) The reliability index β with the required reliability R.

The reliability index β with the required reliability $R = 0.99$ per Equation (3.2) is

$$\beta = \Phi^{-1}(0.99) = 2.32635. \tag{d}$$

(4) Use the modified H-L method to determine the dimension.

Following the procedure and the flowchart in Figure 3.2 discussed above, we can compile a MATLAB program. The program is listed in Appendix B.4 as "M-H-L-program-Example 3.3." The iterative results are listed in Table 3.8.

Table 3.8: The iterative results of Example 3.3 by the modified H-L method

Iterative #	S_y^*	F^*	K_t^*	d^*	$|\Delta d^*|$
1	32.2	28.72	1.9	1.468913	
2	26.20276	32.92314	1.969613	1.775099	0.306186
3	25.437	32.08462	2.286689	1.916347	0.141248
4	25.3237	32.1278	2.286277	1.921748	0.005401
5	25.31118	32.11423	2.286569	1.92194	0.000192
6	25.31073	32.11421	2.286546	1.921946	6.79E-06

According to the result obtained from the program, the mean of the diameter with a reliability 0.99 is

$$\mu_d = 1.922''. \tag{e}$$

Therefore, the diameter of the bar with the required reliability 0.99 under the specified loading will be

$$d = 1.922 \pm 0.005''.$$

∎

Example 3.4
Use the modified H-L method to do Example 3.2.

On the critical cross-section of a beam with a rectangular cross-section is subjected to a bending moment $M = 50.25 \pm 4.16$ (klb.in). The yield strength S_y of this bar's material follows a normal distribution a mean $\mu_{S_y} = 32.2$ (ksi) and a standard deviation $\sigma_{S_y} = 3.63$ (ksi). If the width of the beam is $b = 2.000 \pm 0.010''$. Determine the height h of the bar with a reliability 0.95 when the dimension tolerance is $\pm0.010''$.

Solution:

For this problem, there is not static stress concentration. So we do not need to do the preliminary design.

(1) The limit state function.

The normal stress of the beam caused by the bending moment M is

$$\sigma = \frac{Mh/2}{bh^3/12} = \frac{6M}{bh^2}. \tag{a}$$

The limit state function of the beam is

$$g\left(S_y, M, b, h\right) = S_y - \frac{6M}{bh^2} = \begin{cases} > 0 & \text{Safe} \\ 0 & \text{Limit state} \\ < 0 & \text{Failure.} \end{cases} \tag{b}$$

The bending moment M can be treated as a normal distribution. Its mean and standard deviation can be determined per Equation (1.2). For the dimension, the dimension standard deviation σ_d can be determined per Equation (1.1). In the limit state function, there are four normally distributed variables. Their distribution parameters in the limit state function (b) are listed in Table 3.9.

(2) The reliability index β with the required reliability 0.95.

The reliability index β with the required reliability $R = 0.95$ per Equation (3.2) is

$$\beta = \Phi^{-1}\left(0.0.95\right) = 1.64485. \tag{c}$$

Table 3.9: Distribution parameters for Example 3.4

S_y (ksi)		M (klb.in)		b (in)		h (in)	
μ_{S_y}	σ_{S_y}	μ_M	σ_M	μ_b	σ_b	μ_h	σ_h
32.2	3.63	50.25	1.04	2.000	0.0025	μ_h	0.0025

Table 3.10: The iterative results of Example 3.1 by the modified H-L method

Iterative #	S_y^*	M^*	b^*	h^*	$|\Delta d^*|$
1	32.2	50.25	2	2.163718	
2	26.32887	50.55881	1.981349	2.411446	0.247727
3	26.29544	50.50239	1.984518	2.409705	0.001741
4	26.29542	50.50235	1.984563	2.409678	2.68E-05

(3) Use the modified H-L method to determine the dimension.

We can follow the procedure, and the flowchart in Figure 3.2 discussed above, we can make the MATLAB program. The iterative results are listed in Table 3.10.

According to the result obtained from the program, the mean of the beam heigh with a reliability 0.99 is

$$\mu_h = 2.410''. \tag{d}$$

Therefore, the height of the beam with the required reliability 0.95 under the specified loading will be

$$h = 2.410 \pm 0.010''.$$

$\blacksquare$

3.2.4 DIMENSION DESIGN BY THE MODIFIED R-F METHOD

When the limit state function (3.1) contains at least one non-normal distribution, we need to use the modified R-F method to design the component dimension of a component with the required reliability under specified loading [5]. The R-F method iteratively calculates the reliability index β and then uses converged reliability index β to calculate the reliability, which has been discussed in Section 3.7 of Volume 1 [1] and is also concisely displayed in Appendix A.2 of this book. Since the reliability index β is a known value for a dimension design, we will use the modified R-F method to determine iteratively the mean μ_d of the dimension d. The general procedure of the modified R-F method for dimension design is explained and displayed as follows.

Step 1: Preliminary design for determining K_t for static design or K_f and k_b for fatigue design.

Per Section 3.2.1, we can determine the dimension-dependent parameters K_t for static design or K_f and k_b for fatigue design if necessary.

Step 2: Establish the limit state function.

For a clear description of the modified R-F method procedure, we can rearrange the limit state function (3.1) into the following form of the limit state function.

$$g\left(X_1,\ldots,X_r,\ X_{r+1},\ldots,X_n,d\right)=\begin{cases}>0 & \text{Safe}\\ =0 & \text{Limit state}\\ <0 & \text{Failure,}\end{cases} \qquad (3.18)$$

where the first r random variables are non-normal distributed random variables and the rest $(n-r)$ random variables are normally distributed random variables. The last one is normally distributed dimension with a mean μ_d and a standard deviation σ_d.

Step 3: Calculate the reliability index β.

According to the required reliability of the component, we can determine the reliability index β per Equation (3.2) and is redisplayed here:

$$\beta = \Phi^{-1}(R) = norminv(R). \qquad (3.2)$$

Step 4: Calculate the mean for non-normal distributed random variables.

For non-normally distributed random variables, we can calculate their means based on their type of distributions, which was discussed in Chapter 3 of Volume 1 [1].

Step 5: Pick an initial design point $P^{*0}\left(X_1^{*0},\ldots,X_r^{*0},X_{r+1}^{*0},\ldots,X_n^{*0},d^{*0}\right)$.

The initial design point could be any point, but it must be on the surface of the limit state function. Typically, we can use the means of the first n random variables as the values X_i^{*0} ($i = 1,\ldots,n$) in the initial design point, the value d^{*0} will be determined by the limit state function:

$$\begin{aligned}X_i^{*0} &= \mu_{X_i} \qquad\qquad i = 1,2,\ldots,n\\ g\left(X_1^{*0},\ldots,X_n^{*0},d^{*0}\right) &= 0,\end{aligned} \qquad (3.19)$$

where μ_{X_i} is the mean of every random variable X_i ($i = 1,\ldots,n$). For a non-normally distributed random variable X_i ($i = 1,\ldots,r$), we will use the mean values that have been calculated in Step 4.

When the actual limit state function is provided, we can rearrange the second equation in Equation (3.19) and express d^{*0} by using $X_1^{*0}, X_2^{*0},\ldots$, and X_n^{*0}. Let's use the following equation to represent this:

$$d^{*0} = g_1\left(X_1^{*0}, X_2^{*0},\ldots,X_n^{*0}\right). \qquad (3.20)$$

Step 6: The means and standard deviations of variables at the design point $P^{*0}\left(X_1^{*0}, \ldots, X_n^{*0}, d^{*0}\right)$.

For non-normally distributed random variables, we convert them into equivalent normal distributed random variable [1] per Equation (3.21). For the first r non-normal random variables in the limit state function described in Equation (3.18), we have the following equations:

$$z_{X_i}^{*0} = \Phi^{-1}\left[F_{X_i}\left(X_i^{*0}\right)\right] = norminv\left(F_{X_i}\left(X_i^{*0}\right)\right)$$

$$\sigma_{X_i eq} = \frac{1}{f_{X_i}\left(X_i^{*0}\right)}\phi\left(z_{X_i}^{*0}\right) = \frac{1}{f_{X_i}\left(X_i^{*0}\right)}normpdf\left(z_{X_i}^{*0}\right) \qquad i = 1, 2, \ldots, r \qquad (3.21)$$

$$\mu_{X_i eq} = x_i^{*0} - z_{X_i}^{*0} \times \sigma_{X_i eq},$$

where x_i^{*0} is the value of the non-normally distributed random variable X_i at the design point P^{*0}. $f_{X_i}\left(x_i^{*0}\right)$ and $F_{X_i}\left(x_i^{*0}\right)$ are the PDF and CDF of the non-normally distributed random variable X_i at the design point P^{*0}. $\mu_{X_i eq}$ and $\sigma_{X_i eq}$ are the equivalent mean and the equivalent standard deviation of the equivalent normally distributed random variable.

Now every random variable in the limit state function in Equation (3.18) at the design point P^{*0} is a normally distributed random variable. The mean μ_{eX_i} and standard deviation σ_{eX_i} of these normal distributed random variable at the design point P^{*0} are:

$$\mu_{eX_i} = \begin{cases} \mu_{X_i eq} & i = 1, 2, \ldots, r \\ \mu_{X_i} & i = r + 1, \ldots, n \end{cases} \qquad (3.22)$$

$$\sigma_{eX_i} = \begin{cases} \sigma_{X_i eq} & i = 1, 2, \ldots, r \\ \sigma_{X_i} & i = r + 1, \ldots, n. \end{cases} \qquad (3.23)$$

Step 7: Calculate the new design point $P^{*1}\left(X_1^{*1}, \ldots, X_n^{*1}, d^{*1}\right)$.

For the R-F method, we can calculate the Taylor Series coefficients:

$$G_i|_{P*0} = \sigma_{eX_i} \left.\frac{\partial g\left(X_1, \ldots, X_n, d\right)}{\partial X_i}\right|_{P*0} \qquad i = 1, 2, \ldots, n$$

$$G_d|_{P*0} = \sigma_d \left.\frac{\partial g\left(X_1, \ldots, X_n, d\right)}{\partial d}\right|_{P*0}. \qquad (3.24)$$

Let's use variable G_0 to represent the following equation:

$$G_0 = \sqrt{\sum_{i=1}^{n}\left(G_i|_{P*0}\right)^2 + \left(G_d|_{P*0}\right)^2}. \qquad (3.25)$$

The new design point $P^{*1}\left(X_1^{*1}, \ldots, X_n^{*1}, d^{*1}\right)$ can be calculated by the following equations:

$$X_i^{*1} = \mu_{X_i} + \sigma_{X_i} \times \beta \times \left(\frac{-G_i|_{P*0}}{G_0}\right) \qquad i = 1, \ldots, n \qquad (3.26)$$

d^{*1} will be determined by the limit state function per Equation (3.18) and can be displayed as the following equation:

$$d^{*1} = g_1\left(X_1^{*1}, X_2^{*1}, \ldots, X_n^{*1}\right). \tag{3.27}$$

Step 8: Update the dimension-dependent parameters.

If there are some dimension-dependent parameters such as static stress concentration factor K_t for static loading, the fatigue stress concentration factor K_f and the size factor k_b for cyclic loading, we need to use the new dimensions to update these dimension-dependent parameters. Since the value of any random variable at the design point in the H-L method is determined by this equation,

$$d^{*1} = \mu_d + \sigma_d \times \beta \times \left(\frac{-G_d|_{P*0}}{G_0}\right).$$

Rearranging the above equation, we can get the equation for μ_d:

$$\mu_d = d^{*1} - \sigma_d \times \beta \times \left(\frac{-G_d|_{P*0}}{G_0}\right), \tag{3.28}$$

where $\left(\dfrac{-G_d|_{P*0}}{G_0}\right)$ will be the value calculated in Equations (3.24) and (3.25).

After this μ_d is known, the geometric dimensions for the stress concentration areas are all known. We can update the static stress concentration factor K_t if necessary. Then, use Equations (2.22), (2.23), and (2.24) to update the fatigue stress concentration factor K_f and use Equation (2.17) to update the size modification factor k_b if necessary.

Step 9: Check the convergence condition and the mean μ_d of the dimension d.

d is the dimension in the unit of inch. Therefore, the convergence condition for the dimension d can be:

$$abs\left(d^{*1} - d^{*0}\right) < 0.0001''. \tag{3.29}$$

If the convergence condition (3.29) is not satisfied, we need to update the design point by using the following recurrence of Equation (3.30) and go back to Step 6:

$$\begin{aligned} X_i^{*0} &= X_i^{*1} \qquad i = 1, \ldots, n \\ d^{*0} &= d^{*1}. \end{aligned} \tag{3.30}$$

If the convergence condition (3.29) is satisfied, the μ_d in Equation (3.28) is the mean of the dimension with the required reliability under the specified loading.

Since the modified R-F method is an iterative process, we should use the program for calculation. The program flowchart is shown in Figure 3.3.

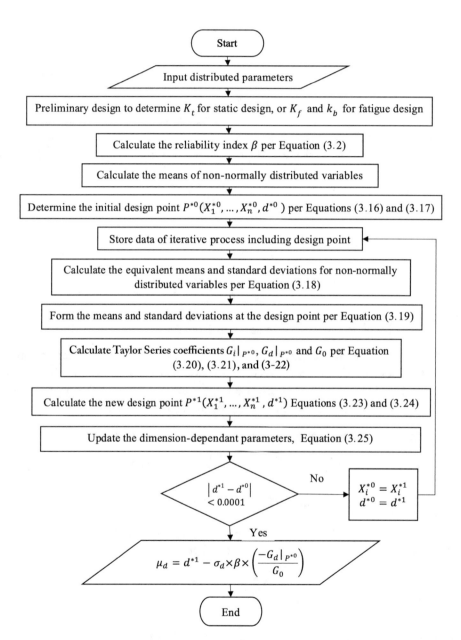

Figure 3.3: The program flowchart for the modified R-F method.

Example 3.5

The critical section of a shaft is in the shoulder section, as shown in Figure 3.4. The shear yield strength S_{sy} (ksi) of a shaft follows a normal distribution with a mean $\mu_{S_{sy}} = 31$ (ksi) and the standard deviation $\sigma_{S_{sy}} = 2.4$ (ksi). The torque applied on the shaft T (klb.in) follows a two-parameter Weibull distribution with the scale parameter $\eta = 20$ and the shape parameter $\beta = 3$. Determine the diameter of the shaft with a reliability 0.99 when the dimension tolerance is ± 0.005.

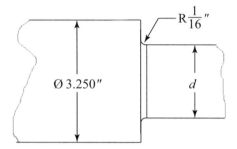

Figure 3.4: Schematic of the segment of a shoulder of a shaft.

Solution:

(1) Preliminary design for determining the static shear stress concentration factor K_{ts}.

This is a static design problem. According to the schematic of the stepped shaft, we can assume that it has a well-rounded fillet. Per Table 3.2, we have the preliminary static stress concentration factor for the shearing loading:

$$K_{ts} = 1.6. \tag{a}$$

(2) The limit state function.

The shear stress of the shaft caused by the torque T is

$$\tau = K_{ts} \frac{T}{\left(\dfrac{\pi}{16}\right) d^3} = K_{ts} \frac{16T}{\pi d^3}. \tag{b}$$

The limit state function of the shaft is

$$g\left(T, S_{ys}, K_{ts}, d\right) = S_{ys} - K_{ts} \frac{16T}{\pi d^3} = \begin{cases} > 0 & \text{Safe} \\ 0 & \text{Limit state} \\ < 0 & \text{Failure.} \end{cases} \tag{c}$$

In the limit state function, there are four random variables. S_{ys}, K_{ts}, and d follow normal distributions. T follows two-parameter Weibull distribution. The standard deviation of normally distributed d can be determined per Equation (1.1). The mean and standard deviation of the static shearing stress concentration factor K_{ts} will be determined per Equation (1.3). Their distribution parameters in the limit state function (c) are listed in Table 3.11. The K_{ts} in this table will be updated in each iterative step by using the new available value μ_d of the shaft diameter.

Table 3.11: Distribution parameters for Example 3.5

S_{ys} (ksi)		T (klb.in) Weibull Distribution		K_{ts}		d (in)	
μ_{S_y}	σ_{S_y}	η	β	$\mu_{K_{ts}}$	$\sigma_{K_{ts}}$	μ_d	σ_d
31	2.4	20	3	1.6	0.08	μ_d	0.00125

(3) The reliability index β of the shaft with a reliability 0.99.

The reliability index β with the required reliability $R = 0.99$ per Equation (3.2) is

$$\beta = \Phi^{-1}(0.99) = 2.326348. \qquad (d)$$

(4) Use the modified R-F method to determine the dimension.

Following the modified R-F method procedure discussed above and the flowchart shown in Figure 3.3, we can make a MATLAB program. The program is listed in Appendix B.5 as "M-R-F program-Example 3.5." The iterative results are listed in Table 3.12.

Table 3.12: The iterative results of Example 3.5 by the modified R-F method

Iterative #	T^*	S_y^*	K_{ts}^*	d^*	$|\Delta d^*|$
1	17.85959	31	1.6	1.674429	
2	23.27808	30.65033	1.607528	1.835977	0.161549
3	23.28036	30.64817	1.918192	1.838956	0.002978
4	23.14829	30.51093	1.92268	1.949733	0.110778
5	23.1735	30.50655	1.93525	1.952054	0.002321
6	23.1677	30.50057	1.93568	1.956264	0.004209
7	23.16837	30.5003	1.936161	1.956433	0.000169
8	23.16816	30.50007	1.936187	1.956594	0.000161
9	23.16818	30.50005	1.936205	1.956604	9.47E-06

According to the result obtained from the program, the mean of the diameter with a reliability 0.99 is

$$\mu_d = 1.957''. \qquad (e)$$

Therefore, the diameter of the shaft with the required reliability 0.99 under the specified loading will be

$$d = 1.957 \pm 0.005''.$$

■

Example 3.6

A machined double-shear pin is under a cyclic shearing loading spectrum. The mean shearing loading can be treated as a constant $V_m = 10.125$ (klb). The shearing loading amplitude V_a can be treated as a normal distribution with a mean $\mu_{V_a} = 8.72$ (klb) and a standard deviation $\sigma_{V_a} = 0.357$ (klb). The number of cycles n_L of this cyclic shearing loading is treated as a constant $n_L = 500000$ (cycles). The ultimate material strength S_u of the pin is 75 (ksi). The three parameters of the material fatigue strength index K_0 on the critical section for a fully reversed bending loading on the standard fatigue specimen are $m = 8.21$, $\mu_{\ln K_0} = 41.738$, and $\sigma_{\ln K_0} = 0.357$. For the material fatigue strength index K_0, the stress unit is ksi. Determine the diameter of the pin with a reliability 0.99 when the dimension tolerance is ± 0.005.

Solution

(1) Preliminary design for determining k_b for fatigue design.

This problem does not have any stress concentration. However, it is a fatigue issue; the size modification factor is a dimension-dependent parameter. The preliminary size modification factor per Equation (3.3) will be

$$k_b = 0.87. \tag{a}$$

(2) The cyclic stress and the component fatigue damage index.

The mean shear stress τ_m and the shear stress amplitude τ_a of the pin due to this cyclic shearing loading are:

$$\tau_m = \frac{V_m/2}{A} = \frac{V_m/2}{\pi d^2/4} = \frac{2V_m}{\pi d^2} \tag{b}$$

$$\tau_a = \frac{V_a/2}{A} = \frac{V_a/2}{\pi d^2/4} = \frac{2V_a}{\pi d^2}. \tag{c}$$

Since this is non-zero-mean cyclic shear stress, the equivalent stress amplitude of a fully reversed cyclic shear stress is:

$$\tau_{a-eq} = \tau_a \frac{S_u}{(S_u - \tau_m)} = \frac{2V_a}{\pi d^2} \frac{S_u}{(S_u - 2V_m/\pi d^2)} = \frac{2V_a S_u}{(\pi d^2 S_u - 2V_m)}. \tag{d}$$

The component fatigue damage index of this pin under this model #3 cyclic shear loading per Equation (2.84) is:

$$D = n_L \left[\frac{2V_a S_u}{(\pi d^2 S_u - 2V_m)} \right]^{8.21}. \tag{e}$$

(3) The limit state function.

The limit state function of the pin under this cyclic shearing loading spectrum per Equation (2.87) is:

$$g(K_0, k_a, k_c, V_a, d) = (k_a k_b k_c)^{8.21} K_0 - n_L \left[\frac{2 V_a S_u}{(\pi d^2 S_u - 2 V_m)} \right]^{8.21} = \begin{cases} > 0 & \text{Safe} \\ 0 & \text{Limit state} \\ < 0 & \text{Failure.} \end{cases}$$

(f)

There are five random variables in the limit state function (f). K_0 is a lognormal distribution. All others are normal distributions. The mean and standard deviation of the surface modification factor k_a can be determined per Equations (2.14), (2.15), and (2.16). The mean and standard deviation of the load modification factor k_c can be determined per Equations (2.18), (2.19), and (2.20). The size modification factor k_b is a deterministic and can be determined per Equation (2.17). It needs to be updated in each iterative step when the μ_d is updated. The dimension d is a normal distribution. Its standard deviation can be calculated per Equation (1.1). Their distribution parameters in the limit state function (f) are listed in Table 3.13.

Table 3.13: The distribution parameters of random variables in Equation (f)

K_0		k_a		k_c		V_a (klb)		d (in)	
$\mu_{\ln K_0}$	$\sigma_{\ln K_0}$	μ_{k_a}	σ_{k_a}	μ_{k_c}	σ_{k_c}	μ_{V_a}	σ_{V_a}	μ_d	σ_d
41.738	0.357	0.8588	0.05153	0.583	0.07171	8.72	0.357	μ_d	0.00125

(4) The reliability index β of the double-shear pin with the reliability 0.99.

The reliability index β with the required reliability $R = 0.99$ per Equation (3.2) is

$$\beta = \Phi^{-1}(0.99) = 2.326348.$$

(g)

(5) Use the modified R-F method to determine the dimension.

We can follow the modified R-F method procedure discussed above and the flowchart shown in Figure 3.4 to compile a MATLAB program. The iterative results are listed in Table 3.14.

According to the result obtained from the program, the mean of the diameter d of the pin with the reliability 0.99 is

$$\mu_d = 0.716''.$$

(h)

Therefore, the diameter of the pin with the required reliability 0.99 under the specified loading will be

$$d = 0.716 \pm 0.005''.$$

Table 3.14: The iterative results of Example 3.6 by the modified R-F method

Iterative #	K_0^*	k_a^*	k_c^*	V_a^*	d^*	$|\Delta d^*|$
1	1.43E+18	0.8588	0.583	8.72	0.687351	
2	1.21E+18	0.841437	0.533467	8.802078	0.728019	0.040669
3	1.23E+18	0.841917	0.53143	8.776322	0.715777	0.012242
4	1.23E+18	0.84222	0.532131	8.795148	0.715286	0.000491
5	1.23E+18	0.842223	0.53219	8.796141	0.715262	2.38E-05

∎

3.2.5   DIMENSION DESIGN BY THE MODIFIED MONTE CARLO METHOD

When a limit state function of a component is established, the Monte Carlo method can be utilized to calculate the reliability of a component under specified loading, which is discussed in Section 3.8 of Volume 1 [1] and is also displayed in Appendix A.3 of this book. The Monte Carlo method can be modified to determine the component dimension with the required reliability under specified loading. The idea is to pick an initial value for the dimension and then use the Monte Carlo method to calculate the reliability of the component by using a new dimension with a small incremental in the dimension such as 0.001″ until the calculated reliability is slightly over the required reliability. Then, this dimension will be the dimension of the component with the required reliability under specified loading.

The general procedure by using the modified Monte Carlo method to conduct the dimension design is explained and listed here.

Step 1: Preliminary design for determining K_t for static design or K_f and k_b for fatigue design.

Per Section 3.2.1, we need to determine the dimension-dependent parameters K_t for static design or K_f and k_b for fatigue design if necessary.

Step 2: Establish the limit state function.

Per the question under consideration, we establish its limit state function. We can use the general limit state function (3.1) and redisplay it here:

$$g(X_1, \ldots, X_n, d) = \begin{cases} > 0 & \text{Safe} \\ = 0 & \text{Limit state} \\ < 0 & \text{Failure,} \end{cases} \tag{3.1}$$

where X_i $(i = 1, 2, \ldots, n)$ is a random variable related to component strength or loading, which could be any type of distributions. d is a normal distribution dimension with a mean μ_d and a standard deviation σ_d.

Step 3: Determine the initial value for the mean μ_d^{*0} of the dimension d.

We can choose any value as the initial value μ_d^{*0} for the mean μ_d of the dimension d. To save computing time, we could use the following approach to determine the initial value μ_d^{*0}.

For a strength-related question, we can assume that all random variables except the material strength variable will be deterministic and are replaced by their means. For a deflection related question, we can assume that all random variables except the material Young's modulus E or shear Young's modulus G will be deterministic and are replaced by their means. Let us assume that X_1 in the limit state function (3.1) is component strength variable such as yield strength, ultimate strength, fatigue strength, Young's modulus E, or shear Young's modulus G. Its CDF is $F_{X_1}(x_1)$. We can use the required reliability R to calculate the point x_1^{*0} for the component strength to satisfy this

$$R = P\left(X_1 > x_1^{*0}\right) = 1 - P\left(X_1 < x_1^{*0}\right) = 1 - F_{X_1}\left(x_1^{*0}\right). \tag{3.31}$$

Rearrange the Equation (3.31), we have

$$x_1^{*0} = F_{X_1}^{-1}\left(1 - R\right), \tag{3.32}$$

where R is the required reliability for the dimension design. $F_{X_1}^{-1}(\cdot)$ is the inverse CDF of component strength X_1. μ_d^{*0} can be determined by following equations:

$$g\left(x_1^{*0}, \mu_{X_2}, \ldots, \mu_{X_n}, \mu_d^{*0}\right) = 0, \tag{3.33}$$

where μ_{X_i} $(i = 2, 3, \ldots, n)$ is the mean of the random variable X_i. In Equation (3.33), μ_d^{*0} is the only unknown variable and can be solved per actual limit state function.

Step 4: Update the μ_d and the dimension-dependent parameters.

The increment in the dimension μ_d will be $0.001''$. So, the recurrence equation is

$$\mu_d = \mu_d^{*0} + 0.001''. \tag{3.34}$$

After this μ_d is known, the geometric dimensions for the stress concentration areas are all known. If necessary, we need to update the dimension-dependent parameters. We can update the static stress concentration factor K_t. Then use Equations (2.22), (2.23), and (2.24) to update the fatigue stress concentration factor K_f and use Equation (2.17) to update the size modification factor k_b.

Step 5: Use the Monte Carlo method to calculate the reliability R^{*1} of the component.

Use μ_d as the mean of the normally distributed dimension d. Then we can use the Monte Carlo method per the limit state function (3.1) to calculate the reliability R^{*1}. Since the limit state function of a component is typically not too complicated, the trial number for the Monte Carlo method can be $N = 15,998,400$.

Step 6: Check the convergence condition.

The convergence condition can be

$$\Delta R = R^{*1} - R > 0.0001. \tag{3.35}$$

If the convergence condition (3.35) is not satisfied, we will update the μ_d^{*0} by the following equation and go back to Step 4.

$$\mu_d^{*0} = \mu_d. \tag{3.36}$$

If the convergence condition (3.35) is satisfied, the μ_d will be the mean of the dimension d. The program flowchart by using the modified Monte Carlo method to determine the dimension with the required reliability under specified loading is shown in Figure 3.5.

Example 3.7
A circular rod is subjected to an axial loading F which follows a uniform distribution between 7.00 (klb) and 9.00 (klb). The rod is made of a ductile material. The yield strength S_y of the rod's material follows a normal distribution with a mean $\mu_{S_y} = 34.5$ (ksi) and a standard deviation $\sigma_{S_y} = 3.12$ (ksi). Determine the diameter of the rod with a reliability 0.99 when the dimension tolerance is ± 0.005.

Solution:

In this example, there is no dimension-dependent parameter.

(1) The limit state function.

The normal stress σ of the rod caused by the axial loading F is

$$\sigma = \frac{F}{A} = \frac{F}{\pi d^2/4} = \frac{4F}{\pi d^2}. \tag{a}$$

The limit state function of the rod is

$$g(S_y, T, d) = S_y - \frac{4F}{\pi d^2} = \begin{cases} > 0 & \text{Safe} \\ 0 & \text{Limit state} \\ < 0 & \text{Failure.} \end{cases} \tag{b}$$

In the limit state function, there are three random variables. S_y and d follow normal distributions. The axial loading F follows a uniform distribution. The standard deviation of normally distributed d can be determined per Equation (1.1). Their distribution parameters in the limit state function are listed in Table 3.15.

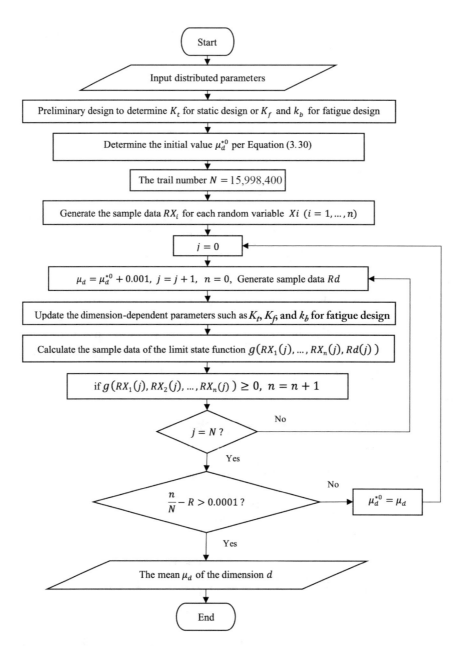

Figure 3.5: The flowchart of the Monte Carlo method for dimension design.

Table 3.15: Distribution parameters for Example 3.7

S_y (ksi) Normal Distribution		F (lb.in) Uniform Distribution		d (in) Normal Distribution	
μ_{S_y}	σ_{S_y}	a	b	μ_d	σ_d
34.5	3.12	7.0	9.0	μ_d	0.00125

(2) Use the modified Monte Carlo method to determine the mean μ_d of the diameter d.

Following the modified Monte Carlo method procedure discussed above and the flowchart shown in Figure 3.5, we can compile a MATLAB program. The program is listed in Appendix B.6 as "M-Monte Carlo program-Example 3.7." The iterative results are listed in Table 3.16.

Table 3.16: The iterative results of Example 3.7 by the modified Monte Carlo method

Iterative #	μ_d^*	R^*	ΔR^*
1	0.61248	0.977524	-0.01248
2	0.61348	0.978904	-0.0111
...	...	...	...
13	0.62448	0.989836	-0.00016
14	0.62548	0.990514	0.000514

According to the result obtained from the program, the mean of the diameter with the reliability 0.99 is

$$\mu_d = 0.626''. \tag{c}$$

Therefore, the diameter of the rod with the required reliability 0.99 under the specified loading will be

$$d = 0.626 \pm 0.005''.$$

∎

Example 3.8
A circular bar with a length $L = 17.000 \pm 0.010''$ is subjected to an axial loading F which follows a normal distribution with a mean $\mu_F = 8.92$ (klb) and a standard deviation $\sigma_F = 0.675$ (klb). The Young's modulus of the bar material follows a normal distribution with the mean $\mu_E = 2.76 \times 10^4$ (ksi) and the standard deviation $\sigma_E = 6.89 \times 10^2$ (ksi). The design specification is that allowable deformation of the entire bar is less than 0.014''. (1) Determine the diameter of the bar with a reliability 0.99 when the dimension tolerance is ± 0.005. (2) Calculate the reliability of

the component with this dimension if the yield strength S_y of the bar's material follows a normal distribution with the mean $\mu_{S_y} = 34.5$ (ksi) and the standard deviation $\sigma_{S_y} = 3.12$ (ksi).

Solution:

In this example, there is no dimension-dependent parameter.

(1) The limit state function for question #1.

The deformation of the bar due to the axial loading F is

$$\delta = \frac{FL}{EA} = \frac{FL}{E\pi d^2/4} = \frac{4FL}{\pi E d^2}. \tag{a}$$

The limit state function of the bar is

$$g(E, F, L, d) = 0.014 - \frac{4FL}{\pi E d^2} = \begin{cases} > 0 & \text{Safe} \\ 0 & \text{Limit state} \\ < 0 & \text{Failure.} \end{cases} \tag{b}$$

In the limit state function, there are four normally distributed random variables. The mean and the standard deviation of length L can be determined per Equation (1.1). The standard deviation of normally distributed d can be determined per Equation (1.1). Their distribution parameters in the limit state function (b) are listed in Table 3.17.

Table 3.17: Distribution parameters for the question #1 in Example 3.8

E (ksi)		F (klb)		L (in)		d (in)	
μ_E	σ_E	μ_F	σ_F	μ_L	σ_L	μ_d	σ_d
2.76×10^4	6.89×10^2	8.92	0.675	17.000	0.0025	μ_d	0.00125

(2) Use the modified Monte Carlo method to determine the mean μ_d of the diameter d.

We can follow the procedure of the modified Monte Carlo method discussed above and the flowchart shown in Figure 3.5 to compile a MATLAB program. The iterative results are listed in Table 3.18.

According to the result obtained from the program, the mean of the diameter of the bar with the required reliability 0.99 is

$$\mu_d = 0.772". \tag{c}$$

Therefore, the diameter of the bar with the required reliability 0.99 under the specified loading will be

$$d = 0.772 \pm 0.005".$$

Table 3.18: The iterative results of Example 3.8 by the modified Monte Carlo Method

Iterative #	μ_d^*	R^*	ΔR^*
1	0.729342	0.789484	-0.20052
2	0.730342	0.79973	-0.19027
...	...	...	...
42	0.770342	0.989449	-0.00055
43	0.771342	0.990444	0.000444

(3) The limit state function for question #2.

The normal stress σ of the bar caused by the axial loading F is

$$\sigma = \frac{F}{A} = \frac{F}{\pi d^2/4} = \frac{4F}{\pi d^2}. \tag{d}$$

The limit state function of the bar is

$$g\left(S_y, F, d\right) = S_y - \frac{4F}{\pi d^2} = \begin{cases} > 0 & \text{Safe} \\ 0 & \text{Limit state} \\ < 0 & \text{Failure.} \end{cases} \tag{e}$$

In the limit state function, there are three normally distributed random variables. Their distribution parameters in the limit state function (e) are listed in Table 3.19.

Table 3.19: Distribution parameters for the question #2 in Example 3.8

S_y (ksi)		F (klb)		d (in)	
μ_E	σ_E	μ_F	σ_F	μ_d	σ_d
34.5	3.12	8.92	0.675	0.772	0.00125

(4) Use the Monte Carlo method to calculate the reliability.

We can follow the procedure of the Monte Carlo method discussed and the flowchart in Appendix A.3 to compile a MATLAB program. The reliability R of the bar with the dimension $d = 0.772 \pm 0.005''$ is:

$$R = \frac{15998345}{15998400} = 0.999997.$$

∎

3.3   DIMENSION OF A COMPONENT WITH REQUIRED RELIABILITY UNDER STATIC LOADING

3.3.1   INTRODUCTION

The limit state functions of a component under different static loading conditions for strength issue and deformation issue have been discussed in Chapter 4 of Volume 1 [1]. After the limit state function of a component under specified static loading is established, the FOSM, modified H-L, modified R-F, and/or the modified Monte Carlo method discussed in Section 3.2 can be used to conduct component dimension design, that is, to determine the dimension with the required reliability.

In this section, we will demonstrate examples to show how to run dimension design with the required reliability under axial loading, direct shearing, torsion, bending moment, and combined loadings.

3.3.2   COMPONENT UNDER STATIC AXIAL LOADING

The limit state function of a component and its reliability calculation under axial loading for strength issue and deformation issue have been discussed in detail in Section 4.6 of Volume 1 [1]. After the limit state function of a component under static axial loading is established, we can run the dimension design with the required reliability. Now we will use examples to show how to conduct component dimension design.

Example 3.9

A rectangular plate is subjected to axial loading. Per design specification, the axial loading follows a normal distribution with a mean $\mu_F = 25.12$ (klb) and a standard deviation $\sigma_F = 3.29$ (klb). The material is ductile, and its yield strength follows a normal distribution with a mean $\mu_{S_y} = 34.5$ (ksi) and a standard deviation $\sigma_{S_y} = 3.12$ (ksi). The Young's modulus of this material follows a normal distribution with a mean $\mu_E = 2.76 \times 10^4$ (ksi) and a standard deviation $\sigma_E = 6.89 \times 10^2$ (ksi). The plate has a thickness $t = 0.375 \pm 0.005''$ and a length $L = 15.250 \pm 0.010''$. Per design specification, the required reliability of this plate is 0.99, and the allowable axial deformation is 0.015''. Use the modified H-L method to determine the height d of the plate with a dimension tolerance $\pm 0.005''$.

Solution:

In this example, there is no dimension-dependent parameter.

(1) The limit state functions.

The deformation of the plate due to the axial loading F is

$$\delta = \frac{FL}{EA} = \frac{FL}{E\,td}. \tag{a}$$

The limit state function of the plate for the deformation issue is

$$g_1(E, F, L, t, d) = 0.015 - \frac{FL}{Etd} = \begin{cases} > 0 & \text{Safe} \\ 0 & \text{Limit state} \\ < 0 & \text{Failure.} \end{cases} \tag{b}$$

The normal stress of the plate due to the axial loading F is

$$\sigma = \frac{F}{A} = \frac{F}{td}. \tag{c}$$

The limit state function of the plate for the strength issues is

$$g_2(S_y, F, t, d) = S_y - \frac{F}{td} = \begin{cases} > 0 & \text{safe} \\ 0 & \text{Limit state} \\ < 0 & \text{Failure.} \end{cases} \tag{d}$$

In these two limit state functions (b) and (d), there are six normally distributed random variables. The mean and the standard deviation of length L and thickness t can be determined per Equation (1.1). The standard deviation of normally distributed d can be determined per Equation (1.1). Their distribution parameters in the two limit state functions (b) and (d) are listed in Table 3.20.

Table 3.20: Distribution parameters for the limit state functions (b) and (d)

E (ksi)		F (klb)		L (in)	
μ_E	σ_E	μ_F	σ_F	μ_L	σ_L
2.76×10^4	6.89×10^2	25.12	3.29	15.25	0.0025
t (ksi)		S_y (klb)		d (in)	
μ_t	σ_t	μ_{S_y}	μ_{S_y}	μ_d	σ_d
0.375	0.00125	34.5	3.12	μ_d	0.00125

(2) Use the modified H-L method to determine the mean μ_d of the height d of the plate.

We can follow the procedure of the modified H-L method discussed in Section 3.2.3 and the flowchart shown in Figure 3.2 to compile a MATLAB program for the limit state function (b) for the deformation issue, and another MATLAB program for the limit state function (d) for the strength issue.

The iterative results for the deformation issue are listed in Table 3.21.

Table 3.21: The iterative results for the limit state function (b)

Iterative #	E^*	F^*	L^*	h^*	d^*	$|\Delta d^*|$
1	27,600	25.12	15.25	0.375	2.467504	
2	27,299.99	32.63592	15.25001	0.374927	3.241643	0.774139
3	27,210.93	32.54067	15.25001	0.374907	3.242937	0.001295
4	27,210.81	32.54054	15.25001	0.374907	3.242937	4.2E-08

According to the result for the strength issue obtained from the program, the mean of the height with a reliability 0.99 is

$$\mu_d = 3.243''. \tag{e}$$

The iterative results for the strength issue are listed in Table 3.22.

Table 3.22: The iterative results for the limit state function (d)

Iterative #	S_y^*	F^*	t^*	d^*	$|\Delta d^*|$
1	34.5	25.12	0.375	1.941643	
2	30.37683	31.41671	0.374939	2.758402	0.81676
3	29.41902	30.58276	0.374934	2.772642	0.014239
4	29.40566	30.56907	0.374936	2.772645	3.74E-06

According to the result for the strength issue obtained from the program, the mean of the height of the plate with a reliability 0.99 is

$$\mu_d = 2.773''. \tag{f}$$

The heigh d of the plate with the required reliability 0.99 under the specified loading will be the larger value of Equations (c) and (f). Therefore, the heigh d of the plate with the required reliability 0.99 under the specified loading and deformation requirement is

$$d = 3.243 \pm 0.005''.$$

■

3.3.3   COMPONENT UNDER STATIC DIRECT SHEARING

The limit state function of a component and its reliability calculation under direct shearing for strength issue have been discussed in detail in Section 4.7 of Volume 1 [1]. After the limit state function of a component under static direct shearing loading is established, we can run the dimension design with the required reliability. Now we will use examples to show how to conduct component dimension design under direct shearing.

Example 3.10

A device consisted of two bars, AC and AB, which are connected through double-double pins, as shown in Figure 3.6. The loading P is 1.8 ± 0.3 (klb). The angle α is $35° \pm 2°$. The pins at point C and B are the same. The shear yield strength S_{sy} of the pins follows a normal distribution with a mean $\mu_{S_{sy}} = 31$ (ksi) and the standard deviation $\sigma_{S_{sy}} = 2.4$ (ksi). Use the modified H-L method to determine the diameter d of the pins at points B and C with a reliability 0.99 when the diameter of the pins has a dimension tolerance $\pm 0.005''$.

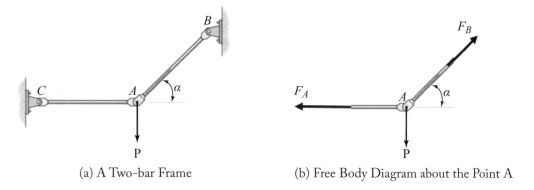

(a) A Two-bar Frame   (b) Free Body Diagram about the Point A

Figure 3.6: The schematic of the device.

Solution:

In this example, there is no dimension-dependent parameter.

(1) Loading analysis.

Based on the Free Body Diagram (FBD) around the point A as shown in Figure 3.4b, the forces F_B and F_A are:

$$F_B = P / \sin(a) \tag{a}$$
$$F_A = P \cos(a) / \sin(a). \tag{b}$$

Since angle a is $35° \pm 2°$, F_A is approximate $1.43P$ and F_B is approximate $1.74P$. Therefore, the pin design for the pins at the point C and point B will be based on the pin at point B since both pins are the same and F_B is larger.

(2) Limit state function of the pin at the point B.

The shear stress of the double-shear pin at point B is

$$\tau = \frac{F_B/2}{A} = \frac{F_B/2}{\pi d^2/4} = \frac{2F_B}{\pi d^2} = \frac{2P}{\pi d^2 \sin(a)}. \qquad (c)$$

The limit state function of the pin at the point B is

$$g\left(S_{sy}, P, a, d\right) = S_{sy} - \frac{2P}{\pi d^2 \sin(a)} = \begin{cases} > 0 & \text{Safe} \\ 0 & \text{Limit state} \\ < 0 & \text{Failure.} \end{cases} \qquad (d)$$

In the limit state function (d), there are four normally distributed random variables. The mean and standard deviation of P can be calculated per Equation (1.2). The mean and standard deviation of angle a in radians can be calculated per Equation (1.1). The standard deviation of the pin diameter d can be calculated per Equation (1.1). Their distribution parameters in the limit state function (d) are listed in Table 3.23.

Table 3.23: Distribution parameters for the limit state function (d) for Example 3.10

S_{ys} (klb)		P (klb)		a (radians)		d (in)	
$\mu_{S_{ys}}$	$\sigma_{S_{ys}}$	μ_P	σ_P	μ_a	σ_a	μ_d	σ_d
31	2.4	1.8	0.075	0.610865	0.008727	μ_d	0.00125

(3) Use the modified H-L method to determine the mean μ_d of the diameter d.

We can follow the procedure of the modified H-L method discussed in Section 3.2.3 and the flowchart shown in Figure 3.2 to compile a MATLAB program for the limit state function (d). The iterative results for this example are listed in Table 3.24.

Table 3.24: The iterative results for Example 3.10

Iterative #	S_y^*	P^*	a^*	d^*	$|\Delta d^*|$
1	31	1.3	0.610865	0.215742	
2	26.59107	1.402672	0.608284	0.242414	0.026671
3	26.2529	1.387884	0.608468	0.242648	0.000235
4	26.24878	1.387767	0.608498	0.242652	3.69E-06

According to the result obtained from the program, the mean of the diameter with a reliability 0.99 is

$$\mu_d = 0.243''. \qquad (e)$$

Therefore, the diameter d of the double-shear pins at the points B and C with the required reliability 0.99 under the specified loading is

$$d = 0.243'' \pm 0.005''. \tag{f}$$

∎

3.3.4 SHAFT UNDER STATIC TORSION LOADING

The limit state function of a component and its reliability calculation under torsion for strength issue and deformation issue have been discussed in detail in Section 4.8 of Volume 1 [1]. After the limit state function of a component under static torsion loading is established, we can run the dimension design with the required reliability. Now we will use examples to show how to conduct component dimension design.

Example 3.11

The solid shaft is subjected to a torque T which follows a uniform distribution between 8.5 (klb.in) and 12.50 (klb.in). The shear yield strength of the shaft material follows a normal distribution with a mean $\mu_{S_{sy}} = 32.2$ (ksi) and the standard deviation $\sigma_{S_{sy}} = 3.63$ (ksi). Use the modified R-F method to design the diameter of the shaft with the required reliability 0.99 when shaft diameter tolerance is ± 0.005.

Solution:

In this example, there is no dimension-dependent parameter.

(1) Limit state function.

The shear stress τ induced by the torque T is

$$\tau = \frac{Td/2}{J} = \frac{Td/2}{\pi d^4/32} = \frac{16T}{\pi d^3}. \tag{a}$$

The limit state function of the shaft for this problem is

$$g\left(T, S_{sy}, d\right) = S_{sy} - \frac{16T}{\pi d^3} = \begin{cases} > 0 & \text{Safe} \\ 0 & \text{Limit state} \\ < 0 & \text{Failure.} \end{cases} \tag{b}$$

In the limit state function (b), there are three random variables. T follows a uniform distribution. The standard deviation of the shaft diameter d can be calculated per Equation (1.1). Their distribution parameters are listed in Table 3.25.

(2) Use the modified R-F method to determine the dimension.

Table 3.25: Distribution parameters for the limit state function (d)

S_{ys} (klb)		T (klb)		d (in)	
Normal Distribution		Uniform Distribution		Normal Distribution	
$\mu_{S_{ys}}$	$\mu_{S_{ys}}$	a	b	μ_d	σ_d
32.2	3.63	8.5	12.5	μ_d	0.00125

We can follow the modified R-F method procedure discussed in Section 3.2.4 and the flowchart shown in Figure 3.3 to compile a MATLAB program. The iterative results are listed in Table 3.26.

Table 3.26: The iterative results of Example 3.11 by the modified R-F method

Iterative #	T^*	S_y^*	d^*	$|\Delta d^*|$
1	10.5	32.2	1.184226	
2	11.57505	30.38601	1.247211	0.062985
3	11.35539	29.91543	1.245735	0.001476
4	11.38899	30.03348	1.245327	0.000408
5	11.38338	30.01924	1.245319	7.52E-06

According to the result obtained from the program, the mean of the diameter with a reliability 0.99 is

$$\mu_d = 1.246''. \tag{d}$$

Therefore, the diameter of the shaft with the required reliability 0.99 under the specified loading will be

$$d = 1.246 \pm 0.005''.$$

∎

Example 3.12

The solid shaft with a constant cross-section and a length $L = 23.75'' \pm 0.032''$ is subjected to a torque T. Per design specification, the torque follows a normal distribution with a mean $\mu_T = 4.5$ (klb.in) and a standard deviation $\sigma_T = 0.39$ (klb.in). The shear yield strength of the shaft material follows a normal distribution with a mean $\mu_{S_{sy}} = 32.2$ (ksi) and a standard deviation $\sigma_{S_{sy}} = 3.63$ (ksi). The shear Young's modulus follows a normal distribution with the mean $\mu_G = 1.117 \times 10^4$ (ksi) and the standard deviation $\sigma_G = 2.793 \times 10^2$ (ksi). The allowable angle of twist of the shaft is 4°. Use the modified Monte Carlo method to design the diameter of the shaft with the required reliability 0.99 when shaft diameter tolerance is ± 0.005.

Solution:

In this example, there is no dimension-dependent parameter.

(1) The limit state functions.

The shear stress τ induced by the torque T is

$$\tau = \frac{Td/2}{J} = \frac{Td/2}{\pi d^4/32} = \frac{16T}{\pi d^3}. \tag{a}$$

The limit state function of the shaft for the strength issue is

$$g\left(S_{sy}, T, d\right) = S_{sy} - \frac{16T}{\pi d^3} = \begin{cases} > 0 & \text{Safe} \\ 0 & \text{Limit state} \\ < 0 & \text{Failure.} \end{cases} \tag{b}$$

The angle of twist of the shaft due to the torque is

$$\Delta\theta = \frac{TL}{GJ} = \frac{TL}{G\pi d^4/32} = \frac{32TL}{G\pi d^4}. \tag{c}$$

The allowable angle of twist $4°$ is equal to 0.069813 radians. The limit state function of the shaft for the deformation issue is

$$g\left(G, T, L, d\right) = 0.069813 - \frac{32TL}{G\pi d^4} = \begin{cases} > 0 & \text{Safe} \\ 0 & \text{Limit state} \\ < 0 & \text{Failure.} \end{cases} \tag{d}$$

In these two limit state functions, there are five normally distributed random variables. The mean and the standard deviation of normally distributed L can be determined per Equation (1.1). The standard deviation of normally distributed d can be determined per Equation (1.1). Their distribution parameters are listed in Table 3.27.

Table 3.27: Distribution parameters for Example 3.12

S_y (ksi)		G (ksi)		T (klb.in)		L (in)		d (in)	
μ_{S_y}	σ_{S_y}	μ_G	σ_G	μ_T	σ_T	μ_L	σ_L	μ_d	σ_d
34.5	3.12	1.117×10^4	2.793×10^2	4.5	0.39	23.75	0.008	μ_d	0.00125

(2) Use the Monte Carlo method to determine the mean μ_d of the diameter d.

Table 3.28: The iterative results of Example 3.12 for the strength issue

Iterative #	μ_d^*	R^*	ΔR^*
1	0.945022	0.969788	-0.02021
2	0.946022	0.971405	-0.01859
...	...	...	...
20	0.964022	0.990068	6.77E-05
21	0.965022	0.99066	0.00066

Following the modified Monte Carlo method procedure discussed in Section 3.2.5 and the flowchart shown in Figure 3.5, we can make MATLAB programs for two limits state functions (c) and (d).

The iterative results for the limit state function (b) are listed in Table 3.28.

According to the result for the strength issue obtained from the program, the mean of the diameter with a reliability 0.99 is

$$\mu_d = 0.966''. \tag{e}$$

The iterative results for the limit state function (d) are listed in Table 3.29.

Table 3.29: The iterative results of Example 3.12 for the deformation issue

Iterative #	μ_d^*	R^*	ΔR^*
1	1.104389	0.765295	-0.22471
2	1.105389	0.778078	-0.21192
...	...	...	...
37	1.140389	0.989331	-0.00067
38	1.141389	0.990534	0.000534

According to the result for the deformation issue obtained from the program, the mean of the diameter with a reliability 0.99 is

$$\mu_d = 1.142''. \tag{f}$$

Therefore, the diameter of the shaft with the required reliability 0.99 under the specified loading and the deformation requirement will be the larger one of Equations (e) and (f). It is

$$d = 1.142 \pm 0.005''.$$

∎

3.3.5   BEAM UNDER STATIC BENDING MOMENT

The limit state function of a component and its reliability calculation under bending moment for strength issue and deflection issue have been discussed in detail in Section 4.9 of Volume 1 [1]. After the limit state function of a component under static bending loading is established, we can run the dimension design with the required reliability. Now we will use examples to show how to conduct component dimension design under bending moment.

Example 3.13

A square cantilever beam with a length $L = 20'' \pm 1/16''$ is subjected to a lateral force F which follows a normal distribution with a mean $\mu_F = 3.675$ (klb) and a standard deviation $\sigma_F = 0.52$ (klb). The yield strength S_y of the beam's material follows a normal distribution with a mean $\mu_{S_y} = 34.5$ (ksi) and a standard deviation $\sigma_{S_y} = 3.12$ (ksi). Use the modified Monte Carlo method to design the side height d of the square beam with the required reliability 0.99 when the side height d has a tolerance ± 0.010.

Solution:

In this example, there is no dimension-dependent parameter.

(1) The limit state function.

   The normal stress of the square beam caused by the lateral force F is

$$\sigma = \frac{FL \times (d/2)}{I} = \frac{FL \times (d/2)}{d^4/12} = \frac{6FL}{d^3}. \tag{a}$$

The limit state function of the beam is

$$g\left(S_y, F, L, d\right) = S_y - \frac{6FL}{d^3} = \begin{cases} > 0 & \text{Safe} \\ 0 & \text{Limit state} \\ < 0 & \text{Failure.} \end{cases} \tag{b}$$

In the limit state function, there are four normally distributed variables. The mean and standard deviation of the length L can be calculated per Equation (1.1). The standard deviation σ_d of the side height, d can be determined per Equation (1.1). Their distribution parameters in the limit state function (b) are listed in Table 3.30.

(2) Use the modified Monte Carlo method to determine the mean μ_d of the height d.

   Following the modified Monte Carlo method procedure discussed in Section 3.2.5 and the flowchart shown in Figure 3.5, we can make MATLAB program for the limits state function (b). The iterative results for the limit state function of this problem are listed in Table 3.31.

Table 3.30: Distribution parameters for Example 3.13

S_y (ksi)		F (klb)		L (in)		d (in)	
μ_{S_y}	σ_{S_y}	μ_F	σ_F	μ_L	σ_L	μ_d	σ_d
34.5	3.12	3.675	0.52	20	0.015625	μ_d	0.0025

Table 3.31: The iterative results of Example 3.13 for the strength issue

Iterative #	μ_d^*	R^*	ΔR^*
1	2.530692	0.929195	-0.0608
2	2.531692	0.930219	-0.05978
...	...	...	...
116	2.645692	0.99001	1.03E-05
117	2.646692	0.990198	0.000198

According to the result obtained from the program, the mean of the side height of the square beam with a reliability 0.99 is

$$\mu_d = 2.647''. \tag{e}$$

Therefore, the side height d of the square beam with the required reliability 0.99 under the specified loading is

$$d = 2.647 \pm 0.005''.$$

■

Example 3.14

A circular simple support beam with a length $L = 24'' \pm 1/16''$ is subjected to a lateral force $F = 4.5 \pm 0.5$ (klb) in the middle of the beam. The yield strength S_y of the beam's material follows a normal distribution with a mean $\mu_{S_y} = 34.5$ (ksi) and a standard deviation $\sigma_{S_y} = 3.12$ (ksi). The Young's modulus E of the beam material follows a normal distribution with a mean $\mu_E = 2.76 \times 10^4$ (ksi) and a standard deviation $\sigma_E = 6.89 \times 10^2$ (ksi). The allowable deflection of the beam is $0.030''$. Use the modified H-L method to design the diameter d of the beam with the required reliability 0.99 when the diameter d has a tolerance ± 0.005.

Solution:

In this example, there is no dimension-dependent parameter.

(1) Limit state functions.

The maximum stress and the maximum deflection of this beam will be in the middle of the beam. The normal stress σ in the middle section of the beam caused by the bending moment is

$$\sigma = \frac{(FL/4) \times (d/2)}{I} = \frac{FLd/8}{\pi d^4/64} = \frac{6FL}{\pi d^3}. \tag{a}$$

The limit state function of the beam for the strength in this problem is

$$g\left(S_y, F, L, d\right) = S_y - \frac{6FL}{\pi d^3} = \begin{cases} > 0 & \text{Safe} \\ 0 & \text{Limit state} \\ < 0 & \text{Failure} \end{cases} \tag{b}$$

The beam deflection in the middle section of the beam in this example is

$$\delta = \frac{FL^3}{48EI} = \frac{FL^3}{48E\left(\pi d^4/64\right)} = \frac{4FL^3}{3E\pi d^3}. \tag{c}$$

The limit state function of the beam for the deflection issue in this problem is

$$g\left(E, F, L, d\right) = 0.030 - \frac{4FL^3}{3E\pi d^3} = \begin{cases} > 0 & \text{Safe} \\ 0 & \text{Limit state} \\ < 0 & \text{Failure.} \end{cases} \tag{d}$$

In these two limit state functions (b) and (d), there are five normally distributed random variables. The mean and the standard deviation of the force F can be calculated per Equation (1.2). The mean and the standard deviation of the length L can be calculated per Equation (1.1). The standard deviation of the beam diameter d can be calculated per Equation (1.1). Their distribution parameters in these two limit state functions are listed in Table 3.32.

Table 3.32: Distribution parameters for the limit state functions (b) and (d)

S_y (ksi)		E (ksi)		F (klb)		L (in)		d (in)	
μ_{S_y}	σ_{S_y}	μ_E	σ_E	μ_F	σ_F	μ_L	σ_L	μ_d	σ_d
34.5	3.12	2.76×10^4	6.89×10^2	4.5	0.125	24	0.015625	μ_d	0.00125

(2) Use the modified H-L method to determine the dimension.

Following the modified H-L method procedure discussed in Section 3.2.3 and the flowchart shown in Figure 3.2, we can compile MATLAB programs for the limit state functions (b) and (d).

Table 3.33: The iterative results for the strength issue in this problem

Iterative #	S_y^*	T^*	L^*	d^*	$|\Delta d^*|$
1	34.5	4.5	24	1.814967	
2	27.56354	4.58536	24.00025	1.968272	0.153306
3	27.44461	4.568076	24.0002	1.96863	0.000358
4	27.44438	4.568041	24.0002	1.968631	2.14E-07

The iterative results for the strength in this problem are listed in Table 3.33.

According to the result for the strength issue obtained from the program, the mean of the diameter with a reliability 0.99 is

$$\mu_d = 1.969''. \tag{e}$$

The iterative results for the deformation issue in this problem are listed in Table 3.34.

Table 3.34: The iterative results for the strength issue in this problem

Iterative #	E^*	T^*	L^*	d^*	$|\Delta d^*|$
1	27600	4.5	24	3.171039	
2	26,530.6	4.715882	24.0019	3.263924	0.092885
3	26,480.34	4.707326	24.00191	3.264013	8.92E-05

According to the result for the deformation issue obtained from the program, the mean of the diameter with a reliability 0.99 is

$$\mu_d = 3.265''. \tag{f}$$

The diameter of the beam with the required reliability 0.99 under the specified loading and deformation requirement will be the larger one of Equations (e) and (f). Therefore, the diameter of the beam is

$$d = 3.265 \pm 0.005''.$$

∎

3.3.6 COMPONENT UNDER STATIC COMBINED LOADING

The limit state function of a component and its reliability calculation under combined loading for strength issue have been discussed in detail in Section 4.10 of Volume 1 [1]. After the limit state function of a component under static combined loading is established, we can run the dimension design with the required reliability. Now we will use examples to show how to conduct component dimension design under combined loading.

Example 3.15

On the critical section of a circular shaft, the resultant internal torsion and an internal resultant bending moment of a shaft are $T = 2.5 \pm 0.18$ (klb.in) and $M = 4.6 \pm 0.34$ (klb.in), as shown in Figure 3.7. The yield strength S_y of the shaft's material follows a normal distribution with the mean $\mu_{S_y} = 34.5$ (ksi) and the standard deviation $\sigma_{S_y} = 3.12$ (ksi). Use the distortion energy theory with the modified Monte Carlo method to design the diameter d of the shaft with the required reliability 0.99 when the diameter d has a tolerance ± 0.005.

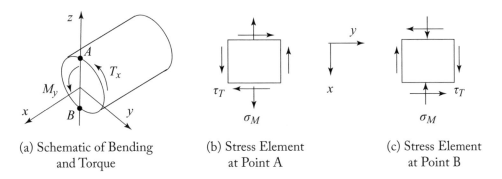

(a) Schematic of Bending and Torque

(b) Stress Element at Point A

(c) Stress Element at Point B

Figure 3.7: Schematic of a segment of a shaft under combined stress.

Solution:

In this example, there is no dimension-dependent parameter.

(1) The Von-Mises stress.

As shown in Figure 3.7, the critical points on this critical section will be the points A and B because there are the maximum values of bending stress. Stress elements at points A and B are shown in Figures 3.7b,c, where σ_M and τ_T are the bending stress due to the bending moment M_y and the shear stress due to the torque T_x, respectively. The Von Mises stress [1, 2] at points A and B for the loading case in this example are the same. The point A is used to run the calculation. At point A, we have:

$$\sigma_x = \sigma_M = \frac{32M_y}{\pi d^3}, \quad \sigma_y = \sigma_z = 0, \quad \tau_{xy} = \tau_T = -\frac{16T_x}{\pi d^3}, \quad \tau_{yz} = \tau_{zx} = 0. \tag{a}$$

The Von Mises in this case is:

$$\sigma_{von} = \sqrt{\left(\frac{32M_y}{\pi d^3}\right)^2 + 3\left(-\frac{16T_x}{\pi d^3}\right)^2} = \frac{16}{\pi d^3}\sqrt{4M_y^2 + 3T_x^2}. \tag{b}$$

(2) The limit state function.

The limit state function of this example by using the distortion energy theory [1] is:

$$g\left(S_y, T_x, M_y, d,\right) = S_y - \frac{16}{\pi d^3} \sqrt{4M_y^2 + 3T_x^2} = \begin{cases} > 0 & \text{Safe} \\ 0 & \text{Limit state} \\ < 0 & \text{Failure.} \end{cases} \qquad (c)$$

There are four normally distributed random variables in this limit state function. The mean and standard deviations of the bending moment and the torque can be calculated per Equation (1.2). The standard deviation of the diameter d can be calculated per Equation (1.1). Their distribution parameters for this example are listed in Table 3.35.

Table 3.35: The distribution parameters for the limit state function (c)

S_y (psi)		T_x (klb.in)		M_y (klb.in)		d (in)	
μ_{S_y}	σ_{S_y}	μ_{T_x}	σ_{T_x}	μ_{M_y}	σ_{M_y}	μ_d	σ_d
34.5	3.12	2.5	0.045	4.6	0.085	μ_d	0.00125

(3) Use the modified H-L method to determine the dimension.

Following the modified H-L method procedure discussed in Section 3.2.3 and the flowchart shown in Figure 3.2, we can make a MATLAB program for the limit state function (c).

The iterative results for this problem are listed in Table 3.36.

Table 3.36: The iterative results for the strength issue in this problem

Iterative #	S_y^*	T_x^*	M_y^*	d^*	$|\Delta d^*|$
1	34.5	2.5	4.6	1.144977	
2	27.35034	2.503722	4.632582	1.239629	0.094652
3	27.30947	2.502936	4.625846	1.23973	0.000102
4	27.30941	2.502938	4.625835	1.239731	1.2E-07

According to the result obtained from the program, the mean of the diameter with the reliability 0.99 is

$$\mu_d = 1.240''. \qquad (d)$$

Therefore, the diameter of the shaft with the required reliability 0.99 under the specified loading is

$$d = 1.240 \pm 0.005''.$$

■

Example 3.16

Schematic of a thin-cylindrical vessel is depicted in Figure 3.8. The vessel has an inner diameter $d = 50'' \pm 0.125''$. The internal pressure is $p = 350 \pm 30$ (psi). The vessel material is ductile. The yield strength S_y of this material follow a normal distribution with a mean $\mu_{S_y} = 34,500$ (psi) and a standard deviation $\sigma_{S_y} = 3120$ (psi). Use the maximum shear stress (MSS) theory with the modified H-L method to design the thickness t of the vessel with a reliability 0.999 when the thickness d has a tolerance $\pm 0.030''$.

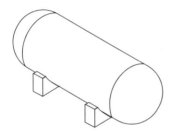

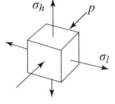

(a) Schematic of a Thin-Wall Vessel   (b) Stress Element of a Critical Point

Figure 3.8: Schematic of a thin-wall cylindrical vessel.

Solution:

In this example, there is no dimension-dependent parameter.

(1) The maximum shear stress.

Stress element of this vessel at the critical point is shown in Figure 3.8b, where p is the internal pressure, σ_l is the longitudinal normal stress and σ_h is the normal stress in the hoop direction. σ_l and σ_h can be calculated by using the following equation:

$$\sigma_h = \frac{pD}{2d}, \qquad \sigma_l = \frac{pD}{4d}, \tag{a}$$

where D is the inner diameter of the vessel, and d is the wall thickness.

Since σ_h, σ_l, and $-p$ are three principal stresses and are arranged as $\sigma_h > \sigma_l > -p$ in this case, the maximum shear stress [1, 2] in this case will be:

$$\tau_{max} = \frac{\sigma_h + p}{2} = p\left(\frac{D}{4d} + \frac{1}{2}\right). \tag{b}$$

(2) The limit state function of the vessel.

The limit state function of this vessel by using the maximum shear stress theory [1] is

$$g\left(S_y, p, D, d\right) = \frac{S_y}{2} - p\left(\frac{D}{4d} + \frac{1}{2}\right) = \begin{cases} > 0 & \text{Safe} \\ 0 & \text{Limit state} \\ < 0 & \text{Failure.} \end{cases} \quad \text{(c)}$$

There are four normally distributed random variables in this limit state function. The mean and standard deviation of the internal pressure p can be determined per Equation (1.2). The mean and standard deviation of the inner diameter D can be determined per Equation (1.1). The standard deviation of the wall thickness d can be determined per Equation (1.1). Their distribution parameters are listed in Table 3.37.

Table 3.37: The distribution parameters of random variables in Equation (c)

S_y (psi)		p (psi)		D (in)		d (in)	
μ_{S_y}	σ_{S_y}	μ_p	σ_p	μ_D	σ_D	μ_d	σ_d
34,500	3,120	350	7.5	50	0.03125	50	0.0075

(3) Use the modified H-L method to determine the dimension.

Following the modified H-L method procedure discussed in Section 3.2.3 and the flowchart shown in Figure 3.2, we can compile a MATLAB program for the limit state function (c). The iterative results for this problem are listed in Table 3.38.

Table 3.38: The iterative results for Example 3.16

Iterative #	S_y^*	p^*	D^*	d^*	$|\Delta d^*|$
1	34,500	350	50	0.256223	
2	25,543.6	355.1015	50.00061	0.352448	0.096226
3	25,132.85	353.8936	50.00047	0.357054	0.004606
4	25,122.04	353.8485	50.00047	0.357164	0.00011
5	25,121.78	353.8474	50.00047	0.357166	2.68E-06

According to the result obtained from the program, the mean of the thickness with a reliability 0.999 is

$$\mu_d = 0.358''. \quad \text{(d)}$$

Therefore, the thickness of the vessel with the required reliability 0.999 under the specified loading is

$$d = 0.358 \pm 0.030''.$$

■

Example 3.17

A schematic of the critical cross-section of a rectangular column is subjected to a combined loading which can be simplified as a compressive force and bending moment, as shown in Figure 3.9. The compression force F_z is along the z-axis and through the centroid of the cross-section, and the bending moment M_y is about the neutral y-axis. The compression force is $F_z = 87 \pm 8$ (klb). The bending moment is $M_y = 610 \pm 40$ (klb.in). The width b of the column is $b = 3 \pm 0.010''$. The column is made of brittle material. Its ultimate tensile strength S_{ut} follows a normal distribution with a mean $\mu_{S_{ut}} = 22.0$ (ksi) and a standard deviation $\sigma_{S_{ut}} = 1.6$ (ksi). The ultimate compression strength S_{uc} follows a normal distribution with a mean $\mu_{S_{uc}} = 43.0$ (ksi) and standard deviation $\sigma_{S_{uc}} = 3.30$ (ksi). Use the maximum normal stress (MNS) theory with the modified Monte Carlo method to design the height h of the column with a reliability 0.95 when the height h has a tolerance $\pm 0.010''$.

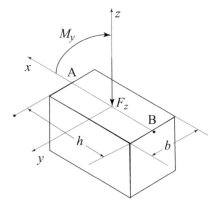

Figure 3.9: Schematic of a segment of a column under compressions and bending.

Solution:

In this example, there is no dimension-dependent parameter.

(1) The maximum tensile normal stress and normal compressive stress.

The maximum tensile stress on the critical cross-section is at the top line A and is equal to

$$\sigma_A = \sigma_{bending} - \sigma_{compression} = \frac{M_y h/2}{bh^3/12} - \frac{F_z}{bh} = \frac{6M_y}{bh^2} - \frac{F_z}{bh}. \tag{a}$$

The maximum compression stress on the critical cross-section as shown in Figure 3.9 is the bottom line B and is equal to

$$\sigma_B = \sigma_{bending} + \sigma_{compression} = \frac{M_y h/2}{bh^3/12} + \frac{F_z}{bh} = \frac{6M_y}{bh^2} + \frac{F_z}{bh}. \tag{b}$$

(2) The limit state functions.

The stress at line A is the maximum tensile stress. The limit state function of the column in line A by using the maximum normal stress theory [1] is

$$g\left(S_{ut}, F_z, M_y, b, h\right) = S_{ut} - \left(\frac{6M_y}{bh^2} - \frac{F_z}{bh}\right) = \begin{cases} > 0 & \text{Safe} \\ 0 & \text{Limit state} \\ < 0 & \text{Failure.} \end{cases} \tag{c}$$

The stress in line B is the maximum compression stress. The limit state function of the column at line B by using the maximum normal stress theory [1] is

$$g\left(S_{uc}, F_z, M_y, b, h\right) = S_{uc} - \left(\frac{6M_y}{bh^2} + \frac{F_z}{bh}\right) = \begin{cases} > 0 & \text{Safe} \\ 0 & \text{Limit state} \\ < 0 & \text{Failure.} \end{cases} \tag{d}$$

There are five random variables in these two limit state functions (c) and (d). The loading F_Z and M_y can be treated as normal distributions. Their distribution parameters can be determined by Equation (1.2). Geometric dimensions b and h can be treated as normal distributions. Their distribution parameters can be determined per Equation (1.1). The distribution parameters for these two limit state functions are listed in Table 3.39.

Table 3.39: The distribution parameters of random variables in Equations (c) and (d)

S_{ut} (ksi)		S_{uc} (ksi)		F_z (klb)		M_y (klb.in)		b (in)		h (in)	
$\mu_{S_{ut}}$	$\sigma_{S_{ut}}$	$\mu_{S_{uc}}$	$\sigma_{S_{uc}}$	μ_{F_z}	σ_{F_z}	μ_{M_y}	σ_{M_y}	μ_b	σ_b	μ_h	σ_h
22	1.6	43	3.3	87	2	610	10	3	0.0025	μ_h	0.0025

(3) Use the modified Monte Carlo method to determine the mean μ_h of the height h.

Following the modified Monte Carlo method discussed in Section 3.2.5 and the flowchart shown in Figure 3.5, we can make MATLAB programs for the limits state functions (c) and (d).

The iterative results for the limit state function (c) are listed in Table 3.40.

Table 3.40: The iterative results of Example 3.17 for the limit state function (c)

Iterative #	μ_h^*	R^*	ΔR^*
1	5.971437	0.94755	-0.00245
2	5.972437	0.947906	-0.00209
...	...	...	...
8	5.978437	0.950026	2.63E-05
9	5.979437	0.950382	0.000382

According to the result for the tension strength obtained from the program, the mean of the height h with a reliability 0.99 is

$$\mu_h = 5.980''. \tag{e}$$

The iterative results for the limit state function (d) are listed in the Table 3.41.

Table 3.41: The iterative results of Example 3.17 for the limit state function (d)

Iterative #	μ_h^*	R^*	ΔR^*
1	6.098313	0.948051	-0.00195
2	6.099313	0.948417	-0.00158
...	...	...	...
6	6.103313	0.949867	-0.00013
7	6.104313	0.950227	0.000227

According to the result for the compression strength obtained from the program, the mean of the height h with a reliability 0.99 is

$$\mu_h = 6.105''. \tag{f}$$

The height h of the rectangular column with the required reliability 0.95 under the specified loading is the larger one of (e) and (f). Therefore, the height h of the rectangular column with the required reliability 0.95 is

$$h = 6.105 \pm 0.010''.$$

■

3.4   DIMENSION OF A COMPONENT WITH REQUIRED RELIABILITY UNDER CYCLIC LOADING SPECTRUM

3.4.1   INTRODUCTION

Two probabilistic fatigue theories, the P-S-N curves approach and the K-D probabilistic fatigue damage model, have been discussed in Chapter 2 for establishing limit state function and calculating the reliability of a component under different cyclic loading spectrum.

Since component stress will be the function of the component dimension, it is an unknown parameter before the dimension design has been completed. Therefore, it is difficult to use the P-N-S curves approach for determining the dimension with the required reliability under cyclic loading spectrum.

The K-D probabilistic fatigue damage model can be used to establish the limit state function of a component under any cyclic loading spectrum. Therefore, it can be used to conduct the component dimension design.

After the limit state function of a component under a cyclic loading spectrum is established, the FOSM, modified H-L method, modified R-F method, and/or modified Monte Carlo method discussed in Section 3.2 can be used to conduct component dimension design, that is, to determine the dimension with the required reliability.

In this section, we first discuss how to conduct component dimension design with infinite life. Then, we will use the K-D probabilistic model to demonstrate examples to show how to run dimension design under axial cyclic loading, cyclic direct shearing loading, cyclic torsion loading, and cyclic bending moment loading. Finally, we will discuss the dimension design of a rotating shaft under cyclic combined loadings.

3.4.2   COMPONENT WITH AN INFINITE FATIGUE LIFE

The limit state function and its reliability calculation of a component under cyclic loading spectrum with an infinite life have been discussed in detail in Section 2.7. After the limit state function of a component for an infinite life is established, we can run the dimension design with the required reliability. Now we will show how to conduct component dimension design with the required reliability and an infinite life under cyclic loading spectrum.

Example 3.18

A machined constant circular bar is subjected a cyclic axial loading. The mean axial loading F_m is equal to 12 (klb). The loading amplitude F_a follows a normal distribution with a mean $\mu_{F_a} = 9.8$ (klb) and a standard deviation $\sigma_{F_a} = 1.1$ (klb). The ultimate material strength is 61.5 (ksi). Its endurance limit S'_e follows a normal distribution with a mean $\mu_{S'_e} = 24.7$ (ksi) and a standard deviation $\sigma_{S'_e} = 2.14$ (ksi), which are based on the fully reversed bending specimen tests. This bar is designed to have an infinite life. Determine the diameter d of the bar with the required reliability 0.99 when it has a dimension tolerance $\pm 0.005''$.

Solution:

(1) Preliminary design for the size modification factor k_b.

Since the loading is a cyclic axial loading, the size modification factor k_b will be equal to 1. So, in this problem, there is no dimension-dependent parameter.

(2) Establish the limit state function of this problem.

The mean stress σ_m and the stress amplitude σ_a of the cyclic axial stress due to the cyclic axial loading can be calculated by the following equations:

$$\sigma_m = \frac{F_m}{\pi d^2/4} = \frac{4F_m}{\pi d^2} \text{ (ksi)} \tag{a}$$

$$\sigma_a = \frac{F_a}{\pi d^2/4} = \frac{4F_a}{\pi d^2} \text{ (ksi).} \tag{b}$$

Since the cyclic axial stress is not a fully reversed cyclic stress, we need to use Equation (2.21) to consider the effect of mean stress and converted it into a fully reversed cyclic stress with an equivalent stress amplitude σ_{a-eq}:

$$\sigma_{a-eq} = \sigma_a \left(\frac{S_{ut}}{S_{ut} - \sigma_m} \right) = \frac{4F_a}{\pi d^2} \left(\frac{S_{ut}}{S_{ut} - \dfrac{4F_m}{\pi d^2}} \right) = \frac{4F_a S_{ut}}{S_{ut}\pi d^2 - 4F_m}. \tag{c}$$

The limit state function of this bar per Equation (2.26) is

$$g\left(S_e', k_a, k_c, F_a, d\right) = k_a k_c S_e' - \left(\frac{4F_a S_{ut}}{S_{ut}\pi d^2 - 4F_m} \right) = \begin{cases} > 0 & \text{Safe} \\ 0 & \text{Limit state} \\ < 0 & \text{Failure.} \end{cases} \tag{d}$$

There are five normally distributed variables in this limit state function. The mean and the standard deviation of the surface finish modification factor k_a can be determined per Equations (2.14), (2.15), and (2.16). The mean and the standard deviation of the load modification factor k_c can be determined per Equations (2.18), (2.19), and (2.20). The mean μ_d and the standard deviation σ_d of the diameter, d can be determined per Equation (1.1). Their distribution parameters for the limit state function (d) are listed in Table 3.42.

(3) Use the modified H-L method to determine the diameter.

All random variable in the limit state function (d) are normal distributions. We can follow the procedure of the modified H-L method discussed in Section 3.2.3 and the program flowchart shown in Figure 3.2 to compile a MATLAB program. The iterative results are listed in Table 3.43.

Table 3.42: The distribution parameters of random variables in Equation (d)

S'_e (ksi)		k_a		k_c		F_a (klb)		d (in)	
$\mu_{S'_e}$	$\sigma_{S'_e}$	μ_{k_a}	σ_{k_a}	μ_{k_c}	σ_{k_c}	μ_{F_a}	σ_{F_a}	μ_d	σ_d
24.7	2.14	0.905	0.0543	0.774	0.1262	9.8	1.1	μ_d	0.00125

Table 3.43: The iterative results of Example 3.18 by the modified H-L method

Iterative #	S'^*_e	k^*_a	k^*_c	F^*_a	d^*	$|\Delta d^*|$
1	24.7	0.905	0.774	11.8	1.056792	
2	22.69742	0.869811	0.551753	10.69245	1.224026	0.167235
3	22.99391	0.876337	0.529924	10.84312	1.241508	0.017482
4	23.05687	0.877242	0.526051	10.87936	1.245111	0.003603
5	23.06867	0.877395	0.52534	10.88653	1.245804	0.000693
6	23.07085	0.877422	0.525209	10.88788	1.245934	0.00013
7	23.07125	0.877427	0.525184	10.88813	1.245958	2.43E-05

According to the result obtained from the program, the mean of the diameter with a reliability 0.99 is

$$\mu_d = 1.246''. \tag{e}$$

Therefore, the diameter of the bar with the required reliability 0.99 under the specified loading is

$$d = 1.246 \pm 0.005''.$$

■

Example 3.19

The critical section for a machined rotating shaft is on the shoulder section, as shown in Figure 3.10. The bending moment M on the shoulder section can be described by a uniform distribution between 1.2 (klb.in) and 1.6 (klb.in). The shaft material's ultimate strength is 61.5 (ksi). Its endurance limit S'_e follows a normal distribution with a mean $\mu_{S'_e} = 24.7$ (ksi) and a standard deviation $\sigma_{S'_e} = 2.14$ (ksi), which are based on the fully reversed bending fatigue specimen tests. This shaft is designed to have an infinite life. Determine the diameter d of the shaft with required reliability 0.99 when it has a dimension tolerance $\pm 0.005''$.

Solution:

(1) Preliminary design for dimension-dependent parameters K_t, K_f, and k_b.

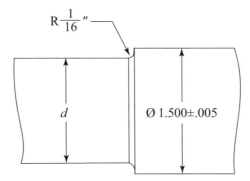

Figure 3.10: Schematic of a shoulder section of a shaft.

According to the preliminary design discussed in Section 3.2.1, we can assume that the fillet, in this case, will be a sharp-fillet. So, the static stress concentration factor K_t is

$$K_t = 2.7. \tag{a}$$

The fatigue stress concentration factor K_f can be calculated per Equations (2.22), (2.23), (2.24), and (2.25):

$$\mu_{K_f} = 2.0337, \qquad \sigma_{K_f} = 0.1627. \tag{b}$$

According to the preliminary design discussed in Section 3.2.1, the size modification factor k_b can be

$$k_b = 0.87. \tag{c}$$

(2) Establish the limit state function of the rotating shaft.

For a rotating shaft, the bending moment will induce a fully reversed cyclic bending stress. Its stress amplitude σ_a of the fully reversed cyclic bending stress will be:

$$\sigma_a = \frac{M \times d/2}{I} = \frac{M \times d/2}{\pi d^4/64} = \frac{32M}{\pi d^3}. \tag{d}$$

Per Equation (2.12), the component endurance limit will be:

$$S_e = k_a k_b k_c S'_e. \tag{e}$$

The surface finish modification factor k_a follows a normal distribution. Its mean and standard deviation can be determined per Equations (2.14), (2.15), and (2.16). k_b is treated as deterministic value and can be calculated per Equation (2.17). It will be updated in each iterative step. The load modification factor k_c will be 1 because the shaft is subjected to cyclic bending stress.

The limit state function of this problem per Equation (2.26) is

$$g\left(M, S'_e, k_a, K_f, d\right) = k_a k_b S'_e - K_f \frac{32M}{\pi d^3} = \begin{cases} > 0 & \text{Safe} \\ 0 & \text{Limit state} \\ < 0 & \text{Failure.} \end{cases} \qquad \text{(f)}$$

There are five random variables in this limit state function. The bending moment M is a uniform distribution. The mean μ_d and the standard deviation σ_d of d can be determined per Equation (1.1). The distribution parameters for the limit state function (f) are listed in Table 3.44. In the table the fatigue stress concentration factor K_f will be updated in each iterative process.

Table 3.44: The distribution parameters for the limit state function (f)

M (klb.in)		S'_e (ksi)		k_a		K_f		d (in)	
a	b	$\mu_{S'_e}$	$\sigma_{S'_e}$	μ_{K_a}	σ_{K_a}	μ_{K_f}	σ_{K_f}	μ_d	σ_d
1.2	1.6	24.7	2.14	0.905	0.0543	2.0337	0.1627	μ_d	0.0125

(3) Use the modified R-F method to determine the diameter d of the shaft.

The limit state function (f) contains four normally distributed random variable and one uniform distribution. We can follow the procedure of the modified R-F method discussed in Section 3.2.4 and the program flowchart shown in Figure 3.3 to compile a MATLAB program. The iterative results are listed in Table 3.45.

Table 3.45: The iterative results of Example 3.19 by the modified R-F method

Iterative #	M^*	S'^*_e	k_a^*	K_f^*	d^*	k_b^*	$|\Delta d^*|$
1	1.4	24.7	0.905	2.0337	1.142484	0.87	
2	1.418724	24.50914	0.901646	3.373608	1.36365	0.842356	0.221167
3	1.495111	23.72887	0.888004	1.598743	1.111104	0.86213	0.252546
4	1.471456	23.66889	0.887261	1.567885	1.09082	0.863932	0.020284
5	1.475578	23.71912	0.888153	1.570136	1.090464	0.863964	0.000357
6	1.474928	23.71568	0.888075	1.570485	1.090456	0.863964	8.25E-06

According to the result obtained from the program, the mean of the diameter with the reliability 0.99 is

$$\mu_d = 1.091''. \qquad \text{(g)}$$

Therefore, the diameter of the shaft with the required reliability 0.99 under the specified cycle bending loading is

$$d = 1.091 \pm 0.005''.$$

■

3.4.3   ROD UNDER CYCLIC AXIAL LOADING SPECTRUM

The limit state function of a rod under any type of cyclic axial loading spectrum can be expressed per Equations (2.87) or (2.88), which have been discussed in Section 2.9.6. After the limit state function of a component under cyclic axial loading is established, we can run the dimension design with the required reliability. In this section, we will show how to determine the dimension of a rod with the required reliability under cyclic axial loading spectrum.

Example 3.20

A machined constant circular bar is subjected to model #2 cyclic axial loading spectrum as listed in Table 3.46. The ultimate material strength S_u is 75 (ksi). The three distribution parameters of material fatigue strength index K_0 for the standard specimen under fully-reversed bending stress are $m = 8.21$, $\mu_{\ln K_0} = 41.738$, and $\sigma_{\ln K_0} = 0.357$. For the material fatigue strength index K_0, the stress unit is ksi. Determine the diameter of the bar with a reliability 0.95 when its dimension tolerance is $\pm 0.005''$.

Table 3.46: The model #2 cyclic axial loading spectrum for Example 3.20

Mean of the Cyclic Axial Loading F_m (klb)	Amplitude of Cyclic Axial Loading F_a (klb)	Number of Cycles n_L (normal distribution)	
		μ_{n_L}	σ_{n_L}
16.78	10.39	1.13×10^5	4.52×10^3

Solution:

For this example, there is no stress concentration area. Because the loading is an axial loading, the size modification factor $k_b = 1$. Therefore, there are no dimension-dependant parameters.

(1) The cyclic axial stress and the component fatigue damage index.

The mean stress σ_m and the stress amplitude σ_a of the bar due to the cyclic axial loading are:

$$\sigma_m = \frac{F_m}{\pi d^2/4} = \frac{4F_m}{\pi d^2}, \qquad \sigma_a = \frac{F_a}{\pi d^2/4} = \frac{4F_a}{\pi d^2}. \tag{a}$$

Since the cyclic stress is a no-zero mean cyclic stress, we need to convert it into a fully reversed cyclic stress per Equation (2.83). The equivalent stress amplitude of this converted fully reversed cyclic stress is:

$$\sigma_{a-eq} = \sigma_a \frac{S_u}{(S_u - \sigma_m)} = \frac{4F_a S_u}{(\pi S_u d^2 - 4F_m)}. \tag{b}$$

The component fatigue damage index D of the bar under model #2 cyclic fatigue spectrum per Equation (2.84) is

$$D = n_L \left(K_f \sigma_a\right)^{8.21} = n_L \left[\frac{4F_a S_u}{(\pi S_u d^2 - 4F_m)}\right]^{8.21}. \tag{c}$$

(2) The limit state function.

The component fatigue strength index K can be calculated per Equation (2.79):

$$K = (k_a k_b k_c)^m K_0. \tag{d}$$

The surface finish modification factor k_a follows a normal distribution. Its mean and standard deviation can be determined per Equations (2.14), (2.15), and (2.16). k_b is 1 for cyclic axial loading. The mean and the standard deviation of the load modification factor k_c can be calculated per Equations (2.18), (2.19), and (2.20).

The limit state function of the bar per Equation (2.87) is

$$g\left(K_0, k_a, k_c, n_L, d\right) = (k_a \, k_c)^m \, K_0 - n_L \left[\frac{4F_a S_u}{(\pi S_u d^2 - 4F_m)}\right]^{8.21} = \begin{cases} > 0 & \text{Safe} \\ 0 & \text{Limit state} \\ < 0 & \text{Failure.} \end{cases} \tag{e}$$

In this limit state function, we have five random variables. K_0 is a log-normal distribution and the rests are normal distributions. The diameter d follows a normal distribution. Its mean and standard deviation can be determined per Equation (1.1). The distribution parameters of these five random variables are listed in Table 3.47.

Table 3.47: The distribution parameters of random variables in Equation (e)

K_0 (lognormal)		k_a		k_c		n_L		d (in)	
$\mu_{\ln K_0}$	$\sigma_{\ln K_0}$	μ_{k_a}	σ_{k_a}	μ_{k_c}	σ_{k_c}	μ_{n_L}	σ_{n_L}	μ_d	σ_d
41.738	0.357	0.8588	0.05153	0.774	0.1262	1.13×10^5	4.52×10^3	μ_d	0.00125

(3) Use the modified Monte Carlo method to determine the diameter d.

In the limit state function (e), there are one lognormal distribution and four normal distribution. We can use the modified Monte Carlo method to conduct this component dimension design, which has been discussed in Section 3.2.5. We can follow the procedure discussed in

Table 3.48: The iterative results of Example 3.20 by the modified Monte Carlo method

Iterative #	μ_d^*	R^*	ΔR^*
1	0.921746	0.705623	-0.28438
2	0.922746	0.71141	-0.27859
...	...	...	...
142	1.062746	0.990046	4.63E-05
143	1.063746	0.990293	0.000293

Section 3.2.5 and the program flowchart shown in Figure 3.5 to compile a MATLAB program. The iterative results are listed in Table 3.48.

According to the result obtained from the program, the mean of the diameter with a reliability 0.99 is

$$\mu_d = 1.064''. \tag{f}$$

Therefore, the diameter of the bar with the required reliability 0.99 under the specified loading will be

$$d = 1.064 \pm 0.005''.$$

∎

Example 3.21

A machined stepped circular bar, as shown in Figure 3.11 is subjected to model #6 cyclic axial loading spectrum listed in Table 3.49. The ultimate material strength S_u is 75 (ksi). The three distribution parameters of material fatigue strength index K_0 for the standard specimen under fully-reversed bending stress are $m = 8.21$, $\mu_{\ln K_0} = 41.738$, and $\sigma_{\ln K_0} = 0.357$. For the material fatigue strength index K_0, the stress unit is ksi. Determine the diameter of the bar with the reliability 0.99 when its dimension tolerance is $\pm 0.005''$.

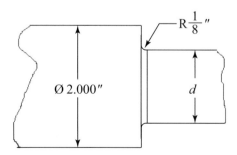

Figure 3.11: Schematic of the segment of a stepped bar.

Table 3.49: Model #6 cyclic axial loading spectrum for Example 3.21

Cyclic Number Level	Number of Cycles n_{Li} (constant)	Fully Reversed Axial Loading Amplitude F_{ai} (klb) (normal distribution)	
		$\mu_{F_{ai}}$	$\sigma_{F_{ai}}$
1	5,000	22.15	3.25
2	200,000	12.45	1.5

Solution:

(1) Preliminary design for determining K_t and K_f.

For the component under cyclic axial loading, the size modification factor k_b will be 1. However, there are two dimension-dependent parameters K_t and K_f in this fatigue problem. According to the schematic of the stepped shaft, we can assume that it has a well-round fillet. Per Table 3.2, we have the preliminary static stress concentration factor

$$K_t = 1.9. \tag{a}$$

K_f follows a normal distribution. Its mean and standard deviation can be calculated per Equations (2.22)–(2.26).

(2) The cyclic axial stress and the component fatigue damage index.

In the cyclic number level #1 with $n_{L1} = 5000$ (cycles), the fully reversed axial stress amplitude σ_{a1} is

$$\sigma_{a1} = K_f \frac{F_{a1}}{\pi d^2/4} = K_f \frac{4F_{a1}}{\pi d^2}, \tag{b}$$

where F_{a1} in klb is the fully reversed axial loading amplitude at the cyclic number level #1.

In the cyclic number level #2 with $n_{L2} = 200{,}000$ (cycles), the fully reversed axial stress amplitude σ_{a2} is

$$\sigma_{a2} = K_f \frac{F_{a2}}{\pi d^2/4} = K_f \frac{4F_{a2}}{\pi d^2}, \tag{c}$$

where F_{a2} in klb is the fully reversed axial loading amplitude at the cyclic number level #2.

The component fatigue damage index D of the bar under this model #6 cyclic axial loading per Equation (2.85) is:

$$D = n_{L1} \left(K_f \frac{4F_{a1}}{\pi d^2} \right)^m + n_{L2} \left(K_f \frac{4F_{a2}}{\pi d^2} \right)^m = \left(\frac{4K_f}{\pi d^2} \right)^m \left[n_{L1} (F_{a1})^m + n_{L2} (F_{a2})^m \right]. \tag{d}$$

(3) The limit state function.

The component fatigue strength index K can be calculated per Equation (2.79):

$$K = (k_a k_b k_c)^m K_0. \tag{e}$$

The surface finish modification factor k_a follows a normal distribution. Its mean and standard deviation can be determined per Equations (2.14), (2.15), and (2.16). k_b is 1 for cyclic axial loading. The mean and the standard deviation of the load modification factor k_c can be calculated per Equations (2.18), (2.19), and (2.20).

The limit state function of the stepped bar in this example per Equation (2.88) is

$$g\left(K_0, k_a, k_c, F_{a1}, F_{a2}, K_f, d\right) = (k_a k_c)^m K_0$$

$$- \left(\frac{4K_f}{\pi d^2}\right)^m \left[n_{L1}\left(F_{a1}\right)^m + n_{L2}\left(F_{a2}\right)^m\right] = \begin{cases} > 0 & \text{Safe} \\ 0 & \text{Limit state} \\ < 0 & \text{Failure.} \end{cases} \tag{f}$$

The diameter d will be treated as a normal distribution. Its mean and standard deviation can be determined per Equation (1.1). There are seven random variables in the limit state function (f). K_0 is a log-normal distribution. The rests are normal distributions. Their distribution parameters in Equation (f) are listed in Table 3.50. In this table, the mean and standard deviation of K_f will be updated in each iterative step by using a new dimension of the bar.

Table 3.50: The distribution parameters of random variables in Equation (d)

K_0 (lognormal)		k_a		k_c		F_{a1} (klb)	
$\mu_{\ln K_0}$	$\sigma_{\ln K_0}$	μ_{k_a}	σ_{k_a}	μ_{k_c}	σ_{k_c}	$\mu_{F_{a1}}$	$\sigma_{F_{a1}}$
41.738	0.357	0.8588	0.05153	0.774	0.1262	22.15	3.25
F_{a2} (klb)		K_f		d (in)			
$\mu_{F_{a2}}$	$\sigma_{F_{a2}}$	μ_{K_f}	σ_{K_f}	μ_d	σ_d		
12.45	1.5	1.6624	0.1330	μ_d	0.00125		

(4) Use the modified R-F method to determine the diameter d.

In the limit state function (f), there are one lognormal distribution and six normal distribution. We will use the modified R-F to conduct this component dimension design, which has been discussed in Section 3.2.4. We can follow the procedure discussed in Section 3.2.4 and the program flowchart shown in Figure 3.3 to compile a MATLAB program. (Note: In this MATLAB program, the subroutine for calculating the stress concentration factor is required.) The iterative results are listed in Table 3.51.

Table 3.51: The iterative results of Example 3.21 by the modified R-F method

Iterative #	K_0^*	k_a^*	k_c^*	F_{a1}^*	F_{a2}^*	K_f^*	d^*	$|\Delta d^*|$
1	1.43E+18	0.8588	0.774	22.15	12.45	1.6624	1.126794	
2	1.28E+18	0.851578	0.725936	24.34357	12.74353	1.687255	1.232736	0.105941
3	1.31E+18	0.854803	0.745879	24.73924	1.190743	1.675838	1.204671	0.028065
4	1.31E+18	0.854718	0.745943	24.73765	1.185953	1.67627	1.204815	0.000144
5	1.31E+18	0.854722	0.745971	24.73794	1.185973	1.676253	1.20479	2.58E-05

According to the result obtained from the program, the mean of the diameter with the reliability 0.99 is

$$\mu_d = 1.205''. \tag{g}$$

Therefore, the diameter of the bar with the required reliability 0.99 under the specified loading will be

$$d = 1.205 \pm 0.005''.$$

■

3.4.4   PIN UNDER CYCLIC DIRECT SHEARING LOADING SPECTRUM

The limit state function of a pin under any cyclic shear loading spectrum can be established per Equations (2.87) or (2.88), which have been discussed in Section 2.9.7. After the limit state function of a component under cyclic direct shearing loading is established, we can run the dimension design with the required reliability. In this section, we will show how to determine the dimension of a pin with the required reliability under cyclic shear loading spectrum.

Example 3.22

A machined single-shear pin is under a cyclic shearing loading spectrum, which is a zero-to-maximum direct shear loading. The maximum shear loading $V_{\max}$ of this cyclic shearing loading can be treated as a constant $V_{\max} = 25.75$ (klb). The number of cycles n_L of this cyclic shearing loading is also treated as a constant $n_L = 500,000$ (cycles). The ultimate material strength S_u is 75 (ksi). The three distribution parameters of material fatigue strength index K_0 for the standard specimen under fully reversed bending stress are $m = 8.21$, $\mu_{\ln K_0} = 41.738$, and $\sigma_{\ln K_0} = 0.357$. For the material fatigue strength index K_0, the stress unit is ksi. Determine the diameter of the pin with a reliability 0.95 when its dimension tolerance is $\pm 0.005''$.

Solution:

(1) Preliminary design for determining the size modification factor k_b.

In this example, the size modification factor k_b is a dimension-dependent parameter. Per the discussion in Section 3.2.1, it will be

$$k_b = 0.87. \tag{a}$$

(2) The cyclic stress and the component fatigue damage index.

The mean shear stress τ_m and the shear stress amplitude τ_a of the pin due to this zero-to-maximum cyclic shearing loading are:

$$\tau_m = \frac{V_m}{A} = \frac{V_{\max}/2}{\pi d^2/4} = \frac{2V_{\max}}{\pi d^2} \tag{b}$$

$$\tau_a = \frac{V_a}{A} = \frac{V_{\max}/2}{\pi d^2/4} = \frac{2V_{\max}}{\pi d^2}. \tag{c}$$

Since this is non-zero-mean cyclic shear stress, the equivalent stress amplitude of a fully reversed cyclic shear stress is:

$$\tau_{a-eq} = \tau_a \frac{S_u}{(S_u - \tau_m)} = \frac{2V_{\max}S_u}{(\pi d^2 S_u - 2V_{\max})}. \tag{d}$$

The component fatigue damage index of this pin under model #1 cyclic shear stress per Equation (2.84) is:

$$D = n_L \left[\frac{2V_{\max}S_u}{(\pi d^2 S_u - 2V_{\max})} \right]^{8.21}. \tag{e}$$

(3) The limit state function.

The component fatigue strength index K can be calculated per Equation (2.79):

$$K = (k_a k_b k_c)^m K_0. \tag{f}$$

The surface finish modification factor k_a follows a normal distribution. Its mean and standard deviation can be determined per Equations (2.14), (2.15), and (2.16). k_b is treated as a deterministic value and can be calculated per Equation (2.17). Its value will be updated in each iterative step by using the newly available diameter of the pin. The mean and standard deviation of the load modification factor k_c can be calculated per Equations (2.18), (2.19), and (2.20).

The limit state function of the pin under model #1 cyclic direct shearing loading spectrum per Equation (2.87) is:

$$g(K_0, k_a, k_c, d) = (k_a k_b k_c)^m K_0 - n_L \left[\frac{2V_{\max}S_u}{(\pi d^2 S_u - 2V_{\max})} \right]^{8.21} = \begin{cases} > 0 & \text{Safe} \\ 0 & \text{Limit state} \\ < 0 & \text{Failure}. \end{cases} \tag{g}$$

Table 3.52: The distribution parameters of random variables in Equation (g)

K_0 (lognormal)		k_a		k_c		d (in)	
$\mu_{\ln K_0}$	$\sigma_{\ln K_0}$	μ_{k_a}	σ_{k_a}	μ_{k_c}	σ_{k_c}	μ_d	σ_d
41.738	0.357	0.8588	0.05153	0.774	0.1262	μ_d	0.00125

There are four random variables in the limit state function (g). K_0 is a log-normal distribution. The rests are normal distributions. The dimension d can be treated as a normal distribution, and its mean and standard deviation can be calculated per Equation (1.1). The distribution parameters in the limit state function (g) are listed in Table 3.52.

(4) Use the modified R-F method to determine the diameter d.

Now, we will use the modified R-F to conduct this component dimension design, which has been discussed in Section 3.2.4. We can follow the procedure discussed in Section 3.2.4 and the program flowchart shown in Figure 3.3 to compile a MATLAB program. The iterative results are listed in Table 3.53.

Table 3.53: The iterative results of Example 3.22 by the modified R-F method

Iterative #	k_0^*	k_a^*	k_c^*	d^*	$|\Delta d^*|$
1	1.43E+18	0.8588	0.774	1.03942	
2	1.3E+18	0.843918	0.674958	1.111562	0.072142
3	1.31E+18	0.845409	0.673574	1.115417	0.003855
4	1.31E+18	0.845455	0.673535	1.115592	0.000175
5	1.31E+18	0.845456	0.673534	1.1156	8.01E-06

According to the result obtained from the program, the mean of the diameter with a reliability 0.99 is

$$\mu_d = 1.116''. \tag{h}$$

Therefore, the diameter of the pin with the required reliability 0.99 under the specified loading will be

$$d = 1.116 \pm 0.005''.$$

■

Example 3.23
A machined double-shear pin is subjected to model #3 cyclic shear loading spectrum as listed in Table 3.54. The ultimate material strength S_u is 75 (ksi). The three distribution parameters of

material fatigue strength index K_0 for the standard specimen under fully reversed bending stress are $m = 8.21$, $\mu_{\ln K_0} = 41.738$, and $\sigma_{\ln K_0} = 0.357$. For the material fatigue strength index K_0, the stress unit is ksi. Determine the diameter of the pin with a reliability 0.99 when its dimension tolerance is $\pm 0.005''$.

Table 3.54: Model #3 cyclic shear loading for Example 3.23

Number of Cycles n_L	Mean of the Cyclic Shear Loading V_m (klb)	Amplitude of Cyclic Shear Loading V_a (klb) (normal distribution)	
		μ_{V_a}	σ_{V_a}
600,000	3.422	4.815	0.6

Solution:

(1) Preliminary design for determining the size modification factor k_b.

In this example, the size modification factor k_b is a dimension-dependent parameter. Per the discussion in Section 3.2.1,

$$k_b = 0.87. \tag{a}$$

(2) The cyclic shear stress and the component fatigue damage index.

The mean shear stress τ_m and the shear stress amplitude τ_a of the double-shear pin in this example are:

$$\tau_m = \frac{V_m/2}{A} = \frac{V_m/2}{\pi d^2/4} = \frac{2V_m}{\pi d^2} \tag{b}$$

$$\tau_a = \frac{V_a/2}{A} = \frac{V_a/2}{\pi d^2/4} = \frac{2V_a}{\pi d^2}. \tag{c}$$

Since this is non-zero-mean cyclic shear stress, the equivalent stress amplitude of a fully reversed cyclic shear stress is:

$$\tau_{a-eq} = \tau_a \frac{S_u}{(S_u - \tau_m)} = \frac{2V_a S_u}{(\pi d^2 S_u - 2V_m)}. \tag{d}$$

The component fatigue damage index of this pin under model #3 cyclic shear stress per Equation (2.84) is:

$$D = n_L \left[\frac{2V_a S_u}{(\pi d^2 S_u - 2V_m)} \right]^{8.21}. \tag{e}$$

(3) The limit state function.

The component fatigue strength index K can be calculated per Equation (2.79):

$$K = (k_a k_b k_c)^m K_0. \tag{f}$$

The surface finish modification factor k_a follows a normal distribution. Its mean and standard deviation can be determined per Equations (2.14), (2.15), and (2.16). k_b is treated as a deterministic value and can be calculated per Equation (2.17). Its value will be updated in each iterative step by using the newly available diameter of the pin. The mean and the standard deviation of the load modification factor k_c can be calculated per Equations (2.18), (2.19), and (2.20).

The limit state function of the double-shear pin under model #1 cyclic shearing loading spectrum per Equation (2.87) is:

$$g(K_0, k_a, k_c, V_a, d) = (k_a k_b k_c)^m K_0 - n_L \left[\frac{2V_a S_{ut}}{(\pi d^2 S_{ut} - 2V_m)} \right]^{8.21} = \begin{cases} > 0 & \text{Safe} \\ 0 & \text{Limit state} \\ < 0 & \text{Failure.} \end{cases} \tag{g}$$

There are five random variables in the limit state function (g). K_0 is a lognormal distribution. The rests are normal distributions. The dimension d can be treated as a normal distribution, and its mean and standard deviation can be calculated per Equation (1.1). The distribution parameters in the limit state function (g) are listed in Table 3.55.

Table 3.55: The distribution parameters of random variables in Equation (g)

K (lognormal)		k_a		k_c		V_a (klb)		d (in)	
$\mu_{\ln K}$	$\sigma_{\ln K}$	μ_{k_a}	σ_{k_a}	μ_{k_c}	σ_{k_c}	μ_{V_a}	σ_{V_a}	μ_d	σ_d
41.738	0.357	0.8588	0.05153	0.774	0.1262	4.815	0.6	μ_d	0.00125

(4) Use the modified Monte Carlo method to determine the diameter d.

We will use the modified Monte Carlo method to conduct this component dimension design, which has been discussed in Section 3.2.5. We can follow the procedure discussed in Section 3.2.5 and the program flowchart shown in Figure 3.5 to compile a MATLAB program. The iterative results are listed in Table 3.56.

According to the result obtained from the program, the mean of the diameter with a reliability 0.99 is

$$\mu_d = 0.550''. \tag{h}$$

Therefore, the diameter of the pin with the required reliability 0.99 under the specified loading will be

$$d = 0.550 \pm 0.005''.$$

∎

Table 3.56: The iterative results of Example 3.23 by the modified Monte Carlo method

Iterative #	μ_d^*	R^*	ΔR^*
1	0.462279	0.806384	-0.18362
2	0.463279	0.811958	-0.17804
...	...	...	...
87	0.548279	0.989843	-0.00016
88	0.549279	0.990191	0.000191

3.4.5   SHAFT UNDER CYCLIC TORSION LOADING SPECTRUM

The limit state function of a shaft under any type of cyclic torsion loading spectrum can be established per Equations (2.87) or (2.88), which have been discussed in Section 2.9.8. After the limit state function of a component under cyclic torsion loading is established, we can run the dimension design with the required reliability. In this section, we will show how to determine the dimension of a shaft with the required reliability under cyclic torsion fatigue spectrum.

Example 3.24

The critical section of a machined shaft with a shoulder is at the shoulder section, as shown in Figure 3.12. It is subjected to model #4 cyclic torsion loading spectrum as listed in Table 3.57. The ultimate material strength S_u is 75 (ksi). The three distribution parameters of material fatigue strength index K_0 for the standard specimen under fully-reversed bending stress are $m = 8.21$, $\mu_{\ln K_0} = 41.738$, and $\sigma_{\ln K_0} = 0.357$. For the material fatigue strength index K_0, the stress unit is ksi. Determine the diameter of the shaft with a reliability 0.99 when its dimension tolerance is $\pm 0.005''$.

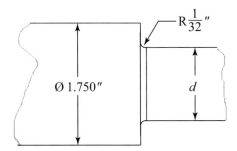

Figure 3.12: Schematic of the segment of a shaft with a shoulder.

Solution:

(1) Preliminary design for determining dimension–dependent parameters.

Table 3.57: The model #4 cyclic fatigue spectrum for Example 3.24

Loading Level #	Number of Cycles n_{Li}	Mean T_{mi} of the Cyclic Torque (klb.in)	Amplitude T_{ai} of the Cyclic Torque (klb.in)
1	500,000	2.25	5.13
2	6,000	2.25	6.42

For the component under cyclic torsion loading, there are three dimension-dependent parameters k_b, K_t, and K_f in this fatigue problem. According to the schematic of the shaft with a shoulder, we can assume that it has a sharp-fillet. The preliminary static stress concentration factor per Table 3.2 will be

$$K_{ts} = 2.2. \tag{a}$$

The fatigue stress concentration factor K_{fs} can be calculated and updated per Equations (2.22)–(2.25).

The preliminary size modification factor k_b per Equation (3.3) will be

$$k_b = 0.87. \tag{b}$$

(2) The cyclic shear stress and the component fatigue damage index.

The mean shear stress τ_{mi}, the shear stress amplitude τ_{ai} and its corresponding equivalent shear stress amplitude τ_{eq-i} of the shaft due to model #4 cyclic torque loading are as follows.

For the loading level #1

$$\tau_{m1} = K_{fs}\frac{T_{m1} \times d/2}{J} = K_{fs}\frac{16T_{m1}}{\pi d^3}. \tag{c}$$

$$\tau_{a1} = K_{fs}\frac{T_{a1} \times d/2}{J} = K_{fs}\frac{16T_{a1}}{\pi d^3} \tag{d}$$

$$\tau_{eq-1} = K_{fs}\tau_{a1}\frac{S_u}{S_u - K_{fs}\tau_{m1}} = K_{fs}\frac{16\tau_{a1}S_u}{\pi d^3 S_u - 16K_{fs}T_{m1}}. \tag{e}$$

For the loading level #2, repeating (c), (d), and (e), we have

$$\tau_{eq-2} = K_{fs}\tau_{a2}\frac{S_u}{S_u - K_{fs}\tau_{m2}} = K_{fs}\frac{16\tau_{a2}S_u}{\pi d^3 S_u - 16K_{fs}T_{m2}}. \tag{f}$$

The component fatigue damage index of this shaft under model #4 cyclic shear stress per Equation (2.85) is:

$$D = n_{L1}\left[K_{fs}\frac{16\tau_{a1}S_u}{\pi d^3 S_u - 16K_{fs}T_{m1}}\right]^{8.21} + n_{L2}\left[K_{fs}\frac{16\tau_{a2}S_u}{\pi d^3 S_u - 16K_{fs}T_{m2}}\right]^{8.21}. \tag{g}$$

(3) The limit state function.

The component fatigue strength index K can be calculated per Equation (2.79).

$$K = (k_a k_b k_c)^m K_0. \tag{h}$$

The surface finish modification factor k_a follows a normal distribution. Its mean and standard deviation can be determined per Equations (2.14), (2.15), and (2.16). k_b is treated as a deterministic value and can be calculated per Equation (2.17). Its value will be updated in each iterative step by using the newly available diameter of the shaft. The mean and the standard deviation of the load modification factor k_c can be calculated per Equations (2.18), (2.19), and (2.20).

The limit state function of the shaft for this example per Equation (2.88) is

$$g\left(K_0, k_a, k_c, K_{fs}, d\right) = (k_a k_b k_c)^m K_0 - n_{L1}\left[K_{fs}\frac{16\tau_{a1}S_u}{\pi d^3 S_u - 16K_{fs}T_{m1}}\right]^{8.21}$$

$$- n_{L2}\left[K_{fs}\frac{16\tau_{a2}S_u}{\pi d^3 S_u - 16K_{fs}T_{m2}}\right]^{8.21} = \begin{cases} > 0 & \text{Safe} \\ 0 & \text{Limit state} \\ < 0 & \text{Failure.} \end{cases} \tag{i}$$

There are five random variables in the limit state function (i). The K_0 is a lognormal distribution. The dimension d can be treated as a normal distribution, and its mean and standard deviation can be calculated per Equation (1.1). The distribution parameters in the limit state function (i) are listed in Table 3.58.

Table 3.58: The distribution parameters of random variables in Equation (i)

K (lognormal)		k_a		k_c		K_{fs}		d (in)	
$\mu_{\ln K}$	$\sigma_{\ln K}$	μ_{k_a}	σ_{k_a}	μ_{k_c}	σ_{k_c}	$\mu_{K_{fs}}$	$\sigma_{K_{fs}}$	μ_d	σ_d
41.738	0.357	0.8588	0.05153	0.774	0.1262	1.6552	0.1324	1.25	0.00125

(4) Use the modified R-F method to determine the diameter d.

In the limit state function (i), there are one lognormal distribution and four normal distribution. Now, we will use the modified R-F to conduct this component dimension design, which has been discussed in Section 3.2.4. We can follow the procedure discussed in Section 3.2.4 and the program flowchart shown in Figure 3.4 to compile a MATLAB program. The iterative progress is listed in Table 3.59.

According to the result obtained from the program, the mean of the diameter with a reliability 0.99 is

$$\mu_d = 1.363''. \tag{j}$$

Table 3.59: The iterative results of Example 3.24 by the modified R-F method

Iterative #	k_0^*	k_a^*	k_c^*	K_{fs}^*	d^*	$|\Delta d^*|$
1	1.43E+18	0.8588	0.774	1.6552	1.365609	
2	1.33E+18	0.854412	0.744798	1.660519	1.365754	0.000145
3	1.33E+18	0.85575	0.753018	1.529347	1.362163	0.003591
4	1.34E+18	0.857546	0.76545	1.526983	1.362098	6.49E-05

Therefore, the diameter of the shaft with the required reliability 0.99 under the specified loading will be

$$d = 1.363 \pm 0.005''.$$

∎

Example 3.25

A machined constant circular shaft is subjected to model #5 cyclic torsion loading spectrum as listed in Table 3.60. The ultimate material strength S_u is 75 (ksi). The three distribution parameters of material fatigue strength index K_0 for the standard specimen under fully-reversed bending stress are $m = 8.21$, $\mu_{\ln K_0} = 41.738$, and $\sigma_{\ln K_0} = 0.357$. For the material fatigue strength index K_0, the stress unit is ksi. Determine the diameter of the shaft with a reliability 0.99 when its dimension tolerance is $\pm 0.005''$.

Table 3.60: The model #5 cyclic fatigue spectrum for Example 3.25

Loading Level#	Torque Mean T_m (klb.in)	Torque Amplitude T_a (klb.in)	Number of Cycles n_L	
			$\mu_{n_{Li}}$	$\sigma_{n_{Li}}$
1	3.5	4.2	300,000	5,000
2	3.5	7.5	4,000	200

Solution:

(1) Preliminary design for determining the size modification factor k_b.

In this example, the size modification factor k_b is the only one dimension-dependent parameter. Per the discussion in Section 3.2.1, it will be:

$$k_b = 0.87. \tag{a}$$

(2) The cyclic shear stress and the component fatigue damage index.

The mean shear stress τ_m, the shear stress amplitude τ_a and its corresponding equivalent shear stress amplitude τ_{eq} of the shaft due to the model #5 cyclic torque loading are as follows.

For the loading level #1,

$$\tau_{m1} = \frac{T_{m1} \times d/2}{J} = \frac{T_{m1} \times d/2}{\pi d^4/32} = \frac{16 T_{m1}}{\pi d^3} \tag{b}$$

$$\tau_{a1} = \frac{T_{a1} \times d/2}{J} = \frac{T_{a1} \times d/2}{\pi d^4/32} = \frac{16 T_{a1}}{\pi d^3} \tag{c}$$

$$\tau_{eq1} = \tau_{a1} \frac{S_u}{S_u - \tau_{m1}} = \frac{16 T_{a1} S_u}{\pi d^3 S_u - 16 T_{m1}}. \tag{d}$$

For the loading level #2, repeating (b), (c), and (d), we have:

$$\tau_{eq2} = \tau_{a2} \frac{S_u}{S_u - \tau_{m2}} = \frac{16 T_{2a} S_u}{\pi d^3 S_u - 16 T_{m2}}. \tag{e}$$

The component fatigue damage index of this shaft under model #5 cyclic shear stress per Equation (2.85) is:

$$D = n_{L1} \left[\frac{16 \tau_{a1} S_u}{\pi d^3 S_u - 16 T_{m1}} \right]^{8.21} + n_{L2} \left[\frac{16 T_{2a} S_u}{\pi d^3 S_u - 16 T_{m2}} \right]^{8.21}. \tag{f}$$

(3) The limit state function.

The component fatigue strength index K can be calculated per Equation (2.79):

$$K = (k_a k_b k_c)^m K_0. \tag{g}$$

The surface finish modification factor k_a follows a normal distribution. Its mean and standard deviation can be determined per Equations (2.14), (2.15), and (2.16). k_b is treated as a deterministic value and can be calculated per Equation (2.17). Its value will be updated in each iterative step by using the newly available diameter of the shaft. The mean and the standard deviation of the load modification factor k_c can be calculated per Equations (2.18), (2.19), and (2.20).

The limit state function of the shaft for this example per Equation (2.87) is:

$$g(K_0, k_a, k_c, n_{L1}, n_{L2}, d) = (k_a k_b k_c)^m K_0 - n_{L1} \left[\frac{16 \tau_{a1} S_{us}}{\pi d^3 S_{us} - 16 T_{m1}} \right]^{8.21}$$

$$- n_{L2} \left[\frac{16 \tau_{a2} S_{us}}{\pi d^3 S_{us} - 16 T_{m2}} \right]^{8.21} = \begin{cases} > 0 & \text{Safe} \\ 0 & \text{Limit state} \\ < 0 & \text{Failure.} \end{cases} \tag{h}$$

There are six random variables in the limit state function (h). K_0 is a lognormal distribution. The rests are normal distributions. The dimension d can be treated as a normal distribution, and its

Table 3.61: The distribution parameters of random variables in Equation (h)

K (lognormal)		k_a		k_c		n_{L1}		n_{L2}		d (in)	
$\mu_{\ln K}$	$\sigma_{\ln K}$	μ_{k_a}	σ_{k_a}	μ_{k_c}	σ_{k_c}	$\mu_{n_{L1}}$	$\sigma_{n_{L1}}$	$\mu_{n_{L2}}$	$\sigma_{n_{L2}}$	μ_d	σ_d
41.738	0.357	0.8588	0.05153	0.774	0.1262	3E5	5000	4000	200	μ_d	1.25E-3

mean and standard deviation can be calculated per Equation (1.1). The distribution parameters in the limit state function (h) are listed in Table 3.61.

(4) Use the modified Monte Carlo method to determine the diameter d.

We will use the modified Monte Carlo method to conduct this component dimension design, which has been discussed in Section 3.2.5. We can follow the procedure discussed in Section 3.2.5 and the program flowchart shown in Figure 3.5 to compile a MATLAB program. The iterative results are listed in Table 3.62.

Table 3.62: The iterative results of Example 3.25 by the modified Monte Carlo method

Iterative #	μ_d^*	R^*	ΔR^*
1	1.160466	0.849113	-0.14089
2	1.161466	0.852414	-0.13759
...	...	...	...
111	1.270466	0.990013	1.28E-05
112	1.271466	0.990263	0.000263

According to the result obtained from the program, the mean of the diameter with a reliability 0.99 is

$$\mu_d = 1.272''. \tag{i}$$

Therefore, the diameter of the shaft with the required reliability 0.99 under the specified loading will be

$$d = 1.272 \pm 0.005''.$$

∎

3.4.6   BEAM UNDER CYCLIC BENDING LOADING SPECTRUM

The limit state function of a beam under any type of cyclic bending loading spectrum can be established per Equations (2.87) or (2.88), which have been discussed in Section 2.9.9. After the limit state function of a component under cyclic bending loading is established, we can run the dimension design with the required reliability. In this section, we will show how to determine the dimension of a beam with the required reliability under cyclic bending loading spectrum.

Example 3.26

The critical section of a machined circular beam is at the shoulder section, as shown in Figure 3.13 and is subjected to model #3 cyclic bending spectrum listed in Table 3.63. The ultimate material strength S_u is 61.5 (ksi). The three distribution parameters of material fatigue strength index K_0 for the standard specimen under fully-reversed bending stress are $m = 6.38$, $\mu_{\ln K_0} = 32.476$, and $\sigma_{\ln K_0} = 0.279$. For the material fatigue strength index K_0, the stress unit is ksi. Determine the diameter of the beam with a reliability 0.99 when its dimension tolerance is $\pm 0.005''$.

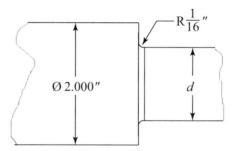

Figure 3.13: Schematic of the segment of a circular beam with a shoulder.

Table 3.63: The model #3 cyclic bending loading spectrum for Example 3.26

Number of Cycles n_L	Bending Moment Mean M_m (klb.in)	Bending Moment Amplitude M_a (klb.in)	
		μ_{M_a}	σ_{M_a}
250,000	0.46	1.08	0.19

Solution:

(1) Preliminary design for determining the dimension-dependent parameters k_b, K_t, and K_f.

Per Section 3.2.1, the preliminary static stress concentration factor per Table 3.2 will be

$$K_t = 2.7. \tag{a}$$

The preliminary size modification factor k_b per Equation (3.3) will be

$$k_b = 0.87. \tag{b}$$

The fatigue stress concentration factor K_f can be calculated and updated per Equations (2.22)–(2.25) when K_t and the fillet radius are known.

(2) The cyclic bending stress and the component fatigue damage index.

The mean bending stress σ_m, the bending stress amplitude σ_a and its corresponding equivalent bending stress amplitude σ_{eq} of the beam due to the model #3 cyclic bending loading listed in the above table are

$$\sigma_m = K_f \frac{M_m \times d/2}{I} = K_f \frac{M_m \times d/2}{\pi d^4/64} = K_f \frac{32 M_m}{\pi d^3} \tag{c}$$

$$\sigma_a = K_f \frac{M_a \times d/2}{I} = K_f \frac{M_a \times d/2}{\pi d^4/64} = K_f \frac{32 M_a}{\pi d^3} \tag{d}$$

$$\sigma_{eq} = K_f \sigma_a \frac{S_u}{S_u - \sigma_m} = \frac{32 M_a S_u K_f}{\pi d^3 S_u - 32 K_f M_m}. \tag{e}$$

The component fatigue damage index of this shaft under model #3 cyclic bending stress per Equation (2.84) is:

$$D = n_L \left(\frac{32 M_a S_u K_f}{\pi d^3 S_u - 32 K_f M_m} \right)^m. \tag{f}$$

(3) The limit state function.

The component fatigue strength index K can be calculated per Equation (2.79):

$$K = (k_a k_b k_c)^m K_0. \tag{g}$$

The surface finish modification factor k_a follows a normal distribution. Its mean and standard deviation can be determined per Equations (2.14), (2.15), and (2.16). k_b is treated as a deterministic value and can be calculated per Equation (2.17). Its value will be updated in each iterative step by using the newly available diameter of the beam. Since this is cyclic bending stress the load modification factor k_c will be 1.

The limit state function of the beam in this example per Equation (2.87) is:

$$g\left(K_0, k_a, K_f, M_a, d\right) = (k_a k_b)^m K_0 - n_L \left(\frac{32 M_a S_u K_f}{\pi d^3 S_u - 32 K_f M_m} \right)^m \tag{3.37}$$

$$= \begin{cases} > 0 & \text{Safe} \\ 0 & \text{Limit state} \\ < 0 & \text{Failure.} \end{cases} \tag{h}$$

There are five random variables in the limit state function (h). K_0 is a lognormal distribution. The rests are normal distributions. The dimensions d can be treated as normal distributions, and its mean and standard deviation can be calculated per Equation (1.1). The distribution parameters in the limit state function (h) are listed in Table 3.64. K_f and k_b will be updated in each iterative step by newly available diameter μ_d.

Table 3.64: The distribution parameters of random variables in Equation (e)

K_0 (lognormal)		k_a		K_f		M_a		d (in)	
$\mu_{\ln K_0}$	$\sigma_{\ln K_0}$	μ_{k_a}	σ_{k_a}	μ_{K_f}	σ_{K_f}	μ_{M_a}	σ_{M_a}	μ_d	σ_d
32.476	0.279	0.9053	0.05432	2.0337	0.1627	1.08	0.19	μ_d	0.00125

(4) Use the modified R-F method to determine the diameter d.

We will use the modified R-F to conduct this component dimension design, which has been discussed in Section 3.2.4. We can follow the procedure discussed in Section 3.2.4 and the program flowchart shown in Figure 3.3 to compile a MATLAB program. The iterative results are listed in Table 3.65.

Table 3.65: The iterative results of Example 3.26 by the modified R-F method

Iterative #	K_0^*	k_a^*	K_f^*	M_a^*	d^*	$|\Delta d^*|$
1	1.32E+14	0.9053	2.0337	1.08	1.111808	
2	1.21E+14	0.892354	2.091971	1.212769	1.171788	0.059979
3	1.2E+14	0.888834	2.110718	1.687193	1.307916	0.136129
4	1.18E+14	0.885946	2.118994	1.68845	1.316975	0.009059
5	1.18E+14	0.884797	2.116866	1.688034	1.317533	0.000558
6	1.18E+14	0.884717	2.116761	1.687974	1.317565	3.2E-05

According to the result obtained from the program, the mean of the diameter with a reliability 0.99 is

$$\mu_d = 1.318''. \tag{i}$$

Therefore, the diameter of the beam with the required reliability 0.99 under the specified loading will be

$$d = 1.318 \pm 0.005''.$$

∎

Example 3.27

The rectangular beam is subjected to model #4 cyclic bending spectrum listed in Table 3.66. The ultimate material strength S_u is 61.5 (ksi). The three distribution parameters of material fatigue strength index K_0 for the standard specimen under fully-reversed bending stress are $m = 6.38$, $\mu_{\ln K_0} = 32.476$, and $\sigma_{\ln K_0} = 0.279$. For the material fatigue strength index K_0, the stress unit is ksi. The width of the beam is $b = 2.000 \pm 0.010''$. Determine the height h of the beam with a reliability 0.99 when its dimension tolerance is $\pm 0.010''$.

Table 3.66: Model #4 cyclic bending loading spectrum for Example 3.27

Stress Level #	Number of Cycles	Cyclic Bending Moment (klb.in)	
		M_{mi}	M_{ai}
1	550,000	38.5	28.92
2	5,000	38.5	42.16

Solution:

(1) Preliminary design for determining the size modification factor k_b.

In this example, there is no stress concentration area. The size modification factor k_b is the only one dimension-dependent parameter. Per the discussion in Section 3.2.1,

$$k_b = 0.87. \tag{a}$$

(2) The cyclic bending stress and the component fatigue damage index.

For the cyclic bending loading level #1, the mean bending stress σ_{m1}, the bending stress amplitude σ_{a1} and its corresponding equivalent bending stress amplitude σ_{eq1} of the beam are

$$\sigma_{m1} = \frac{M_{m1} \times h/2}{I} = \frac{M_{m1} \times h/2}{bh^3/12} = \frac{6M_{m1}}{bh^2} \tag{b}$$

$$\sigma_{a1} = \frac{M_{a1} \times h/2}{I} = \frac{M_{a1} \times h/2}{bh^3/12} = \frac{6M_{a1}}{bh^2} \tag{c}$$

$$\sigma_{eq1} = \sigma_{a1} \frac{S_u}{S_u - \sigma_{m1}} = \frac{6M_{a1}S_u}{bh^2 S_u - 6M_{m1}}. \tag{d}$$

For the cyclic bending loading level #2, repeating (b), (c), and (d), we have

$$\sigma_{eq2} = \frac{6M_{a2}S_u}{bh^2 S_u - 6M_{m2}}. \tag{e}$$

The component fatigue damage index of this beam under model #4 cyclic bending moment per Equation (2.85) is:

$$D = n_{L1} \left(\frac{6M_{a1}S_u}{bh^2 S_u - 6M_{m1}} \right)^m + n_{L2} \left(\frac{6M_{a2}S_u}{bh^2 S_u - 6M_{m2}} \right)^m. \tag{f}$$

(3) The limit state function.

The component fatigue strength index K can be calculated per Equation (2.79).

$$K = (k_a k_b k_c)^m K_0. \tag{g}$$

The surface finish modification factor k_a follows a normal distribution. Its mean and standard deviation can be determined per Equations (2.14), (2.15), and (2.16). k_b is treated as a deterministic value and can be calculated per Equation (2.17). Its value will be updated in each iterative step by using the newly available height h of the beam. Since this is cyclic bending stress the load modification factor k_c will be 1.

The limit state function of the beam under model #4 cyclic bending moment spectrum per Equation (2.88) is:

$$g\left(K_0, k_a, b, h\right) = \left(k_a k_b\right)^m K_0 - n_{L1}\left(\frac{6M_{a1}S_u}{bh^2 S_u - 6M_{m1}}\right)^m$$

$$- n_{L2}\left(\frac{6M_{a2}S_u}{bh^2 S_u - 6M_{m2}}\right)^m = \begin{cases} > 0 & \text{Safe} \\ 0 & \text{Limit state} \\ < 0 & \text{Failure.} \end{cases} \qquad \text{(h)}$$

There are four random variables in the limit state function (h). K_0 is a lognormal distribution. The rests are normal distributions. The dimensions b and h can be treated as normal distributions. Their means and standard deviations can be calculated per Equation (1.1). The distribution parameters in the limit state function (h) are listed in Table 3.67.

Table 3.67: The distribution parameters of random variables in Equation (h)

K_0 (lognormal)		k_a		b (in)		h (in)	
$\mu_{\ln K_0}$	$\sigma_{\ln K_0}$	μ_{k_a}	σ_{k_a}	μ_b	σ_b	μ_h	σ_h
32.476	0.279	0.9053	0.05432	2.000	0.0025	μ_h	0.0025

(5) Use the modified R-F method to determine the height h of the beam.

We will use the modified R-F to conduct this component dimension design, which has been discussed in Section 3.2.4. We can follow the procedure discussed in Section 3.2.4 and the program flowchart shown in Figure 3.3 to compile a MATLAB program. The iterative results are listed in Table 3.68.

According to the result obtained from the program, the mean of the height of the beam with a reliability 0.99 is

$$\mu_h = 2.904''. \qquad \text{(i)}$$

Therefore, the height of the beam with the required reliability 0.99 under the specified loading will be

$$h = 2.904 \pm 0.010''.$$

■

Table 3.68: The iterative results of Example 3.27 by the modified R-F method

Iterative #	K_0^*	k_a^*	b^*	h^*	$|\Delta h^*|$
1	1.32E+14	0.9053	2	2.703691	
2	1.09E+14	0.868503	1.999953	2.777668	0.073977
3	1.11E+14	0.868067	1.999907	2.897338	0.119671
4	1.11E+14	0.867973	1.999954	2.902817	0.005478
5	1.11E+14	0.86797	1.999956	2.903057	0.00024
6	1.11E+14	0.86797	1.999956	2.903068	1.06E-05

3.4.7   COMPONENT UNDER CYCLIC COMBINED LOADING SPECTRUM

The component fatigue issue due to the combined stresses is very complicated because the frequencies of independent loadings might not be in phase. The limit state function of a rotating shaft under cyclic combined bending-torsion loading is provided per Equation (2.95). Now, we will only show one example of a rotating shaft under a cyclic combined bending-torsion loading to determine the diameter with the required reliability.

Example 3.28
The critical section of a solid rotating shaft is at its shoulder section, as shown in Figure 3.14 and is subjected to cyclic combined loading. According to the design specification, the cyclic fatigue loading spectrum can be described by model #4 cyclic combining loads as listed in Table 3.69. The ultimate material strength S_u is 75 (ksi). The three distribution parameters of material fatigue strength index K_0 for the standard specimen under the fully-reversed bending stress are $m = 8.21$, $\mu_{\ln K_0} = 41.738$, and $\sigma_{\ln K_0} = 0.357$. For the material fatigue strength index K_0, the stress unit is ksi. Determine the diameter of the shaft with a reliability 0.99 when its dimension tolerance is $\pm 0.005''$.

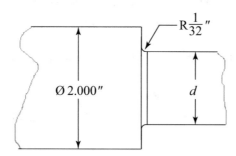

Figure 3.14: Schematic of the segment of a shaft with a shoulder.

Table 3.69: The model #4 cyclic combined loading spectrum for Example 3.28

Level # i	Number of Cycles n_{Li}	Torque T_i (klb.in)	Bending Moment M_i (klb-in)
1	5,500	4.29	5.75
2	580,000	4.29	3.15

Solution:

(1) Preliminary design for determining the dimension-dependent parameters k_b, K_t, and K_f.

For the component under cyclic combined loading with a stress concentration, there are three dimension-dependent parameters k_b, K_f, and K_{fs} in this fatigue problem. According to the schematic of the shaft with a shoulder, we can assume that it has a sharp-fillet. The preliminary static stress concentration factors per Table 3.2 will be

$$K_t = 2.7 \qquad \text{for bending,} \tag{a}$$

$$K_{ts} = 2.2 \qquad \text{for torsion.} \tag{b}$$

The preliminary size modification factor k_b per Equation (3.3) will be

$$k_b = 0.87. \tag{c}$$

The size modification factor k_b will be updated per Equation (2.17). The fatigue stress concentration factor K_f and K_{fs} can be calculated and updated per Equations (2.22)–(2.25) when K_t, K_{ts} and the fillet radius are known.

(2) The cyclic Von Mises stress and the component fatigue damage index.

The mean Von Mises stress σ_{von-mi} per Equation (2.89), the Von Mises stress amplitude σ_{von-ai} per Equation (2.90) and its corresponding equivalent Von Mises stress amplitude σ_{a-eqi} per Equation (2.91) of the rotating shaft in this example are as follows.

For the stress level #1,

$$\sigma_{von-m1} = \sqrt{3}K_{fs}\tau_{T_1} = \sqrt{3}K_{fs}\frac{16T_1}{\pi d^3} \tag{d}$$

$$\sigma_{von-a1} = K_f\sigma_{M_1} = K_f\frac{32M_1}{\pi d^3} \tag{e}$$

$$\sigma_{a-eq1} = \sigma_{von-a1}\frac{S_u}{S_u - \sigma_{von-m}} = \frac{32K_f M_1 S_u}{\pi d^3 S_u - 16K_{fs}\sqrt{3}T_1}. \tag{f}$$

For the stress level #2, we repeat the above calculations:

$$\sigma_{a-eq2} = \frac{32K_f M_2 S_u}{\pi d^3 S_u - 16K_{fs}\sqrt{3}T_2}. \tag{g}$$

The component fatigue damage index of this shaft in this example per Equation (2.94) is:

$$D = n_{L1} \left[\frac{32 K_f M_1 S_u}{\pi d^3 S_u - 16 K_{fs} \sqrt{3} T_1} \right]^{8.21} + n_{L2} \left[\frac{32 K_f M_2 S_u}{\pi d^3 S_u - 16 K_{fs} \sqrt{3} T_2} \right]^{8.21} \tag{h}$$

(3) The limit state function.

The component fatigue strength index K can be calculated per Equation (2.79).

$$K = (k_a k_b k_c)^m K_0. \tag{i}$$

The surface finish modification factor k_a follows a normal distribution. Its mean and standard deviation can be determined per Equations (2.14), (2.15), and (2.16). k_b is treated as a deterministic value and can be calculated per Equation (2.17). The mean and the standard deviation of the load modification factor k_c is assumed to be 1 because this combined cyclic loading is mainly caused by cyclic bending stress.

The limit state function of the rotating shaft in this example per Equation (2.95) is:

$$g\left(K_0, k_a, K_f, K_{fs}, d\right) = (k_a k_b)^m K_0 - n_{L1} \left[\frac{32 K_f M_1 S_u}{\pi d^3 S_u - 16 K_{fs} \sqrt{3} T_1} \right]^{8.21}$$

$$- n_{L2} \left[\frac{32 K_f M_2 S_u}{\pi d^3 S_u - 16 K_{fs} \sqrt{3} T_2} \right]^{8.21} = \begin{cases} > 0 & \text{Safe} \\ 0 & \text{Limit state} \\ < 0 & \text{Failure.} \end{cases} \tag{j}$$

There are five random variables in the limit state function (j). K_0 is a lognormal distribution. The rests are normal distributions. The dimension d can be treated as normal distributions. Its mean and standard deviation can be calculated per Equation (1.1). The distribution parameters in the limit state function (j) are listed in Table 3.70. In the table, K_f and K_{fs} will be updated in each iterative step.

Table 3.70: The distribution parameters of random variables in Equation (j)

K_0 (lognormal)		k_a		K_f		K_{fs}		d (in)	
$\mu_{\ln K_0}$	$\sigma_{\ln K_0}$	μ_{k_a}	σ_{k_a}	μ_{K_f}	σ_{K_f}	$\mu_{K_{fs}}$	$\sigma_{K_{fs}}$	μ_d	σ_d
41.738	0.357	0.9053	0.05432	2.0337	0.1627	1.6552	0.1324	μ_d	0.00125

(4) Use the modified R-F method to determine the diameter d with reliability 0.99.

We will use the modified R-F to conduct this component dimension design, which has been discussed in Section 3.2.4. We can follow the procedure discussed in Section 3.2.4 and the

Table 3.71: The iterative results of Example 3.28 by the modified R-F method

Iterative #	K_0^*	k_a^*	K_f^*	K_{fs}^*	d^*	$|\Delta d^*|$
1	1.43E+18	0.9053	2.0337	1.6552	1.541001	
2	1.34E+18	0.9053	2.0337	1.6552	1.544186	0.003186
3	1.34E+18	0.9053	1.567166	1.751261	1.470725	0.073461
4	1.34E+18	0.9053	1.581338	1.771731	1.473547	0.002822
5	1.34E+18	0.9053	1.581675	1.77091	1.473636	8.86E-05

program flowchart shown in Figure 3.3 to compile a MATLAB program. The iterative results are listed in Table 3.71.

According to the result obtained from the program, the mean of the diameter with a reliability 0.99 is

$$\mu_d = 1.474''. \tag{k}$$

Therefore, the diameter of the shaft with the required reliability 0.99 under the specified loading will be

$$d = 1.474 \pm 0.005''.$$

∎

3.5   SUMMARY

For mechanical component design, we have selected material and have the loadings which are specified by the design specifications. The reliability of a component links the material properties, loading, and dimension together through a limit state function. When the reliability is specified, we can use this limit state function to determine the dimension of the component uniquely.

For mechanical component dimension design, there are several dimensions-dependent parameters. For a component under static loading, the dimension-dependent parameter is the stress concentration factor. For a component under cyclic loading, the dimension-dependent parameters will be the fatigue stress concentration factor and the size modification factor. For mechanical component design, typically, we always have a sketch of the assembly which are based on the required functions of the assembly and components. Therefore we have the rough shape of the component and the possible fillet radius on the stress concentration area if there is a stress concentration. We can conduct preliminary design to select the static stress concentration factor and the size modification factor which has discussed in Section 3.2.1. When the fillet radius in the stress concentration is provided, we can further calculate the fatigue stress concentration factor for fatigue design, which is discussed in Section 2.6. During the iterative process, these dimension-dependent parameters need to be updated in each iterative step by the newly available dimension.

When the limit state function of a component under a specified loading is established, and initial dimension-dependent parameters are selected through the preliminary design, we can use four different computational methods to iteratively determine component dimension with the required reliability under the specified loadings. These four methods are as follows.

- The FOSM method. When all random variables in the limit state function are normal distributions, the FOSM method provides an equation to link the reliability index β with the means and the standard deviations of all normally distributed random variables. In this equation, the only one unknown is the dimension and so can be solved. This is an approximate result. This method has been discussed in Section 3.2.2.

- The modified H-L method. When all random variables in the limit state function are normal distributions, the modified H-L method can be used to determine the component dimension iteratively. The detailed procedure and the program flowchart of the modified H-L method have been discussed in Section 3.2.3.

- The modified R-F method. When there is at least one non-normal distribution in the limit state function, the modified R-F can be used to determine the component dimension iteratively. The detailed procedure and the program flowchart of the modified R-F method have been discussed in Section 3.2.4.

- The modified Monte Carlo Method. For any type of distribution in a limit state function, the modified Monet Carlo method can always be used to determine the component dimension iteratively. This method might take a little longer computing time to get the result when it is compared with the modified H-L and modified R-F methods. The detailed procedure and the program flowchart of the modified Monet Carlo method have been discussed in Section 3.2.5.

The dimension design of the component under static loading is discussed in Section 3.3. Several examples are provided and discussed, including bars under axial loading, pins under direct shearing, shafts under torsion and beams under bending, and components under combined loadings.

It is difficult to use the P-S-N curves approach to conduct the dimension design because stress is the dimension-dependent, and the corresponding P-S-N curves could not be selected. Therefore, the dimension design of component under cyclic loadings is based on the K-D probabilistic fatigue damage model in this chapter and is fully demonstrated and discussed in Section 3.4. Several examples are presented including bars under cyclic axial loadings, pins under cyclic direct shearing loading, shafts under cyclic torsion loadings, beam under cyclic bending loadings, rotating shaft under cyclic combined-bending-torsion loading.

3.6 REFERENCES

[1] Le, Xiaobin, *Reliability-Based Mechanical Design, Volume 1: Component under Static Load*, Morgan & Claypool Publishers, San Rafael, CA, 2020. 121, 123, 124, 125, 129, 131, 136, 137, 138, 145, 152, 154, 157, 161, 164, 165, 166, 167, 168, 170

[2] Budynas, R. G. and Nisbett, J. K., *Shigley's Mechanical Engineering Design*, 10th ed., Mc-Graw Hill Education, New York, 2014. 122, 165, 167

[3] Le, Xiaobin, The research on reliability fatigue design of mechanical components, Doctoral dissertation, Shanghai Jiao Tong University, Shanghai, 1993. 129

[4] Zong, W. H. and Le, Xiaobin, *Probabilistic Design Method of Mechanical Components*, Shanghai Jiao Tong University Publisher, Shanghai, China, September 1995.

[5] Le, Xiaobin and Johan, R., A Probabilistic Approach for Determining the Component's Dimension under Fatigue Loadings, *ASME International Design Engineering Technical Conferences (IDETC)*, San Diego, CA, August 31–September 2, 2009. DOI: 10.1115/detc2009-86334. 129, 136

[6] Le, Xiaobin, A probabilistic fatigue damage model for describing the entire set of fatigue test data of the same material, *ASME International Mechanical Engineering Congress and Exposition, IMECE–10224*, Salt Lake City, UT, November 8–14, 2019.

3.7 EXERCISES

3.1. Describe and explain one example in your design where the shape of the component has roughly defined the sketch of assembly.

3.2. Describe and explain one example where the radius of the fillet in the stress concentration area is determined by the purchased component.

3.3. Describe and explain one example where the radius of the fillet in the stress concentration area can be treated as a well-rounded fillet.

3.4. A rectangular bar is subjected to an axial loading $F_a = 3.1 \pm 0.60$ (klb). The width of the bar b is $1.500 \pm 0.005''$. The yield strength S_y of the bar of a ductile material follows a normal distribution with a mean $\mu_{S_y} = 34.5$ (ksi) and a standard deviation $\mu_{S_y} = 3.12$ (ksi). Use the FORM method to determine the height h of the bar with a reliability 0.95 when the height h has a dimension tolerance $\pm 0.005''$.

3.5. A constant round bar is subjected to an axial loading $F_a = 2.1 \pm 0.20$ (klb). The yield strength S_y of the bar of a ductile material follows a normal distribution with a mean $\mu_{S_y} = 34.5$ (ksi) and the standard deviation $\mu_{S_y} = 3.12$ (ksi). Use the FOSM method to determine the diameter of the bar with a reliability 0.99 when the diameter has a dimension tolerance ±0.005″.

3.6. Use the modified H-L method to do Problem 3.5.

3.7. A bar connected to the supporter at point A is subjected to two concentrated loads F_B and F_C as shown in Figure 3.15. The axial loads are: $F_B = 1500 \pm 120$ (lb) and $F_C = 1000 \pm 90$ (lb). The length of the AB segment $L_1 = 8.00 \pm 0.003″$ and the length of BC segment $L_2 = 10,00 \pm 0.003″$. The Young's modulus E of the bar material follows a normal distribution with a mean $\mu_E = 2.73 \times 10^7$ (psi) and a standard deviation $\sigma_E = 1.30 \times 10^6$ (psi). The maximum allowable deflection of the bar is 0.008″. Use the modified H-L method to determine the diameter of the bar with a reliability 0.99 when its dimension tolerance is ±0.005″.

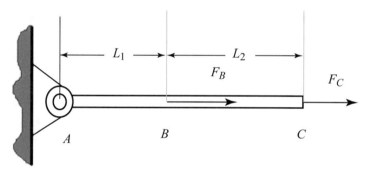

Figure 3.15: A bar under two axial loading.

3.8. Use the modified Monte Carlo method to do Problem 3.7.

3.9. The pin at point A, as shown in Figure 3.15, is a double shear pin. The shear yield strength S_{sy} of the pin of a ductile material follows a normal distribution with a mean $\mu_{S_{sy}} = 32.2$ (ksi) and a standard deviation $\sigma_{S_{sy}} = 3.63$ (ksi). Use the H-L method to determine the diameter of the pin with a reliability 0.99 when its dimension tolerance is ±0.005″.

3.10. Use the modified Monte Carlo method to do Problem 3.9.

3.11. The double shear pin is subjected to a shearing force V (klb) which can be described by a lognormal distribution with a log mean $\mu_{\ln V} = 0.721$ and a log standard deviation $\sigma_{\ln V} = 0.137$. The shear yield strength S_{sy} of the pin of a ductile material follows a normal distribution with a mean $\mu_{S_{sy}} = 32.2$ (ksi) and the standard deviation $\sigma_{S_{sy}} = 3.63$ (ksi). Use the modified R-F method to determine the diameter of the pin with a reliability 0.99 when its dimension tolerance is ±0.005″.

3.12. Use the modified Monte Carlo method to do Problem 3.11.

3.13. A constant cross-section shaft is subjected to a torque T. The torque T (lb.in) can be described by a lognormal distribution with a mean $\mu_{\ln T} = 7.76$ and a standard deviation $\sigma_{\ln T} = 0.194$. The shear yield strength S_{sy} of the shaft of a ductile material follows a normal distribution with a mean $\mu_{S_{sy}} = 32,200$ (psi) and the standard deviation $\sigma_{S_{sy}} = 3630$ (psi). Use the modified R-F method to determine the diameter of the shaft with a reliability 0.99 if its dimension tolerance is $\pm 0.005''$.

3.14. Schematic of a segment of a shaft at its critical cross-section as shown in Figure 3.16 is subjected to a torque $T = 1350 \pm 95$ (lb.in). The fillet radius r and the larger diameter d_2 are $r = 1/32''$, and $d_2 = 1.500 \pm 0.005''$. The shear yield strength S_{sy} of the shaft of a ductile material follows a normal distribution with a mean $\mu_{S_{sy}} = 32,200$ (psi) and the standard deviation $\sigma_{S_{sy}} = 3630$ (psi). Use the modified H-L method to determine the diameter d_2 with a reliability 0.99 when its dimension tolerance is $\pm 0.005''$.

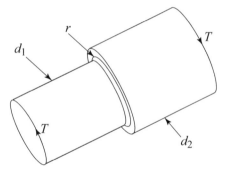

Figure 3.16: Schematic of a segment of a shaft.

3.15. Use a modified Monte Carlo method to do Problem 3.14.

3.16. A constant cross-section shaft with a length $L = 15.00 \pm 0.032''$ is subjected to a pair of opposite torques at both ends $T = 1800 \pm 120$ (lb.in). The shear Young's modulus follows a normal distribution with a mean $\mu_G = 1.117 \times 10^7$ (psi) and a standard deviation $\sigma_G = 2.793 \times 10^5$. If the design requirement is the angle of twist between two ends is less than $1°$, use the modified H-L method to determine the diameter of the shaft with a reliability 0.99 if its dimension tolerance is $\pm 0.005''$.

3.17. Use the modified Monte Carlo method to do Problem 3.16.

3.18. A simple support circular beam is subjected to a concentrated force in the middle, as shown in Figure 3.17. The concentrated force is $P = 1500 \pm 180$ (lb). The span of the beam is $L = 22 \pm 0.065''$. The yield strength S_y of the material follows a normal distribution with the mean $\mu_{S_y} = 34500$ (psi) and the standard deviation $\sigma_{S_y} = 3120$ (psi).

Use the modified H-L method to determine the diameter of the beam with a reliability 0.99 if its dimension tolerance is ±0.005″.

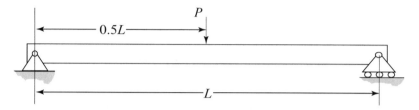

Figure 3.17: A simple support beam.

3.19. Use the modified Monte Carlo method to do Problem 3.18.

3.20. A simple support beam, as shown in Figure 3.18, is subjected to a uniform distributed loading on the beam and a concentrated loading in the middle of the beam. The span of the beam is $L = 15.00 \pm 0.032″$. The beam has a rectangular shape with a width $b = 1.00 \pm 0.010″$. The uniform distributed loading is $w = 100$ (lb/in). The concentrated force in the middle is $P = 1500 \pm 180$ (lb). The yield strength S_y of the material follows a normal distribution with the mean $\mu_{S_y} = 34,500$ (psi) and the standard deviation $\sigma_{S_y} = 3120$ (psi). Use the modified H-L method to determine the height of the beam with a reliability 0.99 if its dimension tolerance is ±0.010″.

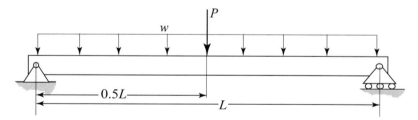

Figure 3.18: A simple support beam.

3.21. Use the modified Monte Carlo method to do Problem 3.20.

3.22. A circular cantilever beam as shown in Figure 3.19 is subjected to a concentrated force at the free end $P = 150 \pm 80$ (lb). The length of the beam is $L = 20.0 \pm 0.032″$. The Young's modulus of the beam material follows a normal distribution with the mean $\mu_E = 2.76 \times 10^7$ (psi) and the standard deviation $\sigma_E = 6.89 \times 10^5$ (psi). The maximum allowable deflection of the beam is $\Delta = 0.022″$. Use the modified H-L method to determine the diameter of the beam with a reliability 0.99 if its dimension tolerance is ±0.005″.

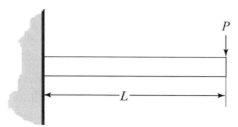

Figure 3.19: A cantilever beam.

3.23. Use the modified Monte Carlo method to do Problem 3.22.

3.24. The critical section of a constant circular shaft is subjected to a torque $T = 3000 \pm$ 150 (lb.in) and a bending moment $M = 9000 \pm 600$ (lb.in). The yield strength S_y of the material follows a normal distribution with the mean $\mu_{S_y} = 34500$ (psi) and the standard deviation $\sigma_{S_y} = 3120$ (psi). The diameter of the shaft has a dimension tolerance $\pm 0.005''$. The required reliability of the shaft is 0.99.

    (a) Use the modified H-L method to determine the diameter of the shaft by using the MSS stress theory.

    (b) Use the modified H-L method to determine the diameter of the shaft by using the DE theory.

    (c) Use the modified Monte Carlo method to determine the diameter of the shaft by using the MSS stress theory.

3.25. A thin-wall cylindrical vessel has an inner diameter $d = 40.0 \pm 0.125''$. The internal pressure of the fluid is $p = 300 \pm 50$ (psi). The yield strength S_y of the material follows a normal distribution with the mean $\mu_{S_y} = 34,500$ (psi) and the standard deviation $\sigma_{S_y} = 3120$ (psi). The thickness the thick vessel has a dimension tolerance $\pm 0.015''$. Use the modified H-L method to determine the thickness of the thin vessel with the reliability 0.999 by using the MSS stress theory.

3.26. Use the modified Monte Carlo method to do Problem 3.25.

3.27. A machined constant circular bar is subjected a cyclic axial loading. The mean axial loading F_m is a constant $F_m = 4.9$ (klb). The loading amplitude F_a is a constant $F_a = 3.84$ (klb). The ultimate material strength is 61.5 (ksi). Its endurance limit S'_e follows a normal distribution with a mean $\mu_{S'_e} = 24.7$ (ksi) and a standard deviation $\sigma_{S'_e} = 2.14$ (ksi), which are based on the fully reversed bending specimen tests. This bar is designed to have an infinite life. Use the modified H-L method to determine the diameter d of the bar with the required reliability 0.99 when it has a dimension tolerance $\pm 0.005''$.

3.28. Use the modified Monte Carlo method to do Problem 3.27.

3.29. The critical section of a machined rotating shaft is at the shoulder section, as shown in Figure 3.20. The bending moment M on the shoulder section follows a normal distribution with a mean $\mu_M = 1.5$ (klb.in) and a standard deviation $\sigma_M = 0.25$ (klb.in). The shaft material's ultimate strength is 61.5 (ksi). Its endurance limit S'_e follows a normal distribution with a mean $\mu_{S'_e} = 24.7$ (ksi) and a standard deviation $\sigma_{S'_e} = 2.14$ (ksi), which are based on the fully reversed bending fatigue specimen tests. This shaft is designed to have an infinite life. Determine the diameter d of the shaft with the required reliability 0.99 when it has a dimension tolerance $\pm 0.005''$.

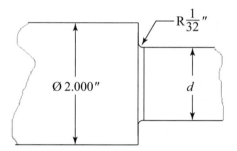

Figure 3.20: Schematic of a segment of a shaft.

3.30. Use a modified Monte Carlo method to do Problem 3.29.

3.31. A machined constant circular bar is subjected to cyclic axial loading spectrum. According to the design specification, the mean of the cyclic axial loading is $F_m = 14.21$ (klb). The amplitude F_a of the cyclic axial loading follows a uniform distribution between 8.25 (klb) and 12.25 (klb). The number of cycles of this cyclic axial loading is $n_L = 450{,}000$ (cycles). The ultimate material strength S_u is 75 (ksi). The three distribution parameters of material fatigue strength index K_0 for the standard specimen under fully-reversed bending stress are $m = 8.21$, $\mu_{\ln K_0} = 41.738$, and $\sigma_{\ln K_0} = 0.357$. For the material fatigue strength index K_0, the stress unit is ksi. Use the modified R-F method to determine the diameter of the bar with the reliability 0.95 when its dimension tolerance is $\pm 0.005''$.

3.32. Use the modified Monte Carlo method to do Problem 3.31.

3.33. A machined stepped circular bar, as shown in Figure 3.20 is subjected to cyclic axial loading spectrum. According to the design specification, the mean F_m and amplitude F_a of the cyclic loading are $F_m = 10.68$ (klb) and $F_a = 7.82$ (klb). The number of cycles of this cyclic axial loading follows a normal distribution with a mean $\mu_{n_L} = 350{,}000$ (cycles) and a standard deviation $\sigma_{n_L} = 14{,}000$ (cycles). The ultimate material strength S_u is

75 (ksi). The three distribution parameters of material fatigue strength index K_0 for the standard specimen under fully-reversed bending stress are $m = 8.21$, $\mu_{\ln K_0} = 41.738$, and $\sigma_{\ln K_0} = 0.357$. For the material fatigue strength index K_0, the stress unit is ksi. Use the modified R-F method to determine the diameter of the bar with the required reliability 0.99 when its dimension tolerance is $\pm 0.005''$.

3.34. Use the modified Monte Carlo method to do Problem 3.33.

3.35. A machined single-shear pin is under a cyclic shearing loading spectrum. The mean of the cyclic shearing loading is $V_m = 12.7$ (klb). The amplitude of the cyclic shearing loading follows a normal distribution with a mean $\mu_{V_a} = 8.39$ (klb) and a standard deviation $\sigma_{V_a} = 1.35$ (klb). The number of cycles n_L of this cyclic shearing loading is a constant $n_L = 500,000$ (cycles). The ultimate material strength S_u is 75 (ksi). The three distribution parameters of material fatigue strength index K_0 for the standard specimen under fully-reversed bending stress are $m = 8.21$, $\mu_{\ln K_0} = 41.738$, and $\sigma_{\ln K_0} = 0.357$. For the material fatigue strength index K_0, the stress unit is ksi. Use the modified R-F method to determine the diameter of the pin with the required reliability 0.95 when its dimension tolerance is $\pm 0.005''$.

3.36. Use the modified Monte Carlo method to do Problem 3.35.

3.37. A machined double-shear pin is subjected to a cyclic shear loading spectrum. According to the design specification, the cyclic shear loading spectrum is listed in Table 3.72. The ultimate material strength S_u is 75 (ksi). The three distribution parameters of material fatigue strength index K_0 for the standard specimen under fully reversed bending stress are $m = 8.21$, $\mu_{\ln K_0} = 41.738$, and $\sigma_{\ln K_0} = 0.357$. For the material fatigue strength index K_0, the stress unit is ksi. Determine the diameter of the pin with a reliability 0.99 when its dimension tolerance is $\pm 0.005''$.

Table 3.72: The cyclic shear loading for Problem 3.37

Loading Level #	Number of Cycles n_L	Mean of the Cyclic Shear Loading V_m (klb)	Amplitude of the Cyclic Shear Loading V_a (klb)
1	4,000	3.422	6.251
2	500,000	3.422	4.815

3.38. Use the modified Monte Carlo method to do Problem 3.37.

3.39. The critical section of a machined shaft with a shoulder is at the shoulder section, as shown in Figure 3.20. It is subjected to a cyclic torque loading spectrum. According to the design specification, the cyclic torsion loading spectrum is listed in Table 3.73. The ultimate material strength S_u is 75 (ksi). The three distribution parameters of material fatigue strength index K_0 for the standard specimen under fully-reversed bending stress are $m = 8.21$, $\mu_{\ln K_0} = 41.738$, and $\sigma_{\ln K_0} = 0.357$. For the material fatigue strength index K_0, the stress unit is ksi. Determine the diameter of the shaft with the required reliability 0.99 when its dimension tolerance is $\pm 0.005''$.

Table 3.73: The cyclic fatigue spectrum for example Problem 3.39

Loading	Number of Cycles n_{Li}		Mean T_{mi} of the Cyclic	Amplitude T_{ai} of the
Level #	$\mu_{n_{Li}}$	$\sigma_{n_{Li}}$	Torque (klb.in)	Cyclic Torque (klb.in)
1	450,000	26,000	1.89	4.32
2	6,500	500	1.89	6.09

3.40. Use the modified Monte Carlo method to do Problem 3.39.

3.41. A machined constant circular shaft is subjected to a cyclic torque loading spectrum. According to the design specification, the cyclic torsion loading spectrum is listed in Table 3.74. The ultimate material strength S_u is 75 (ksi). The three distribution parameters of material fatigue strength index K_0 for the standard specimen under fully reversed bending stress are $m = 8.21$, $\mu_{\ln K_0} = 41.738$, and $\sigma_{\ln K_0} = 0.357$. For the material fatigue strength index K_0, the stress unit is ksi. Use the modified R-F method to determine the diameter of the shaft with a reliability 0.99 when its dimension tolerance is $\pm 0.005''$.

Table 3.74: The cyclic fatigue spectrum for example Problem 3.41

Loading Level#	Number of Cycles n_{Li}	Torque Mean T_m (klb.in)	Torque Amplitude T_a (klb.in)	
			μ_{T_a}	σ_{T_a}
1	4,800,000	3.5	4.2	0.32
2	6,000	3.5	7.5	0.86

3.42. Use the modified Monte Carlo method to do Problem 3.41.

3.43. The constant solid rotating shaft is subjected to cyclic combined bending-torque loading. According to the design specification, the combined cyclic loadings are listed in Table 3.75. The number of cycles of this combined loading is $n_L = 520,000$ (cycles). The ultimate material strength S_u is 75 (ksi). The three distribution parameters of material fatigue strength index K_0 for the standard specimen under fully reversed bending stress are $m = 8.21$, $\mu_{\ln K_0} = 41.738$, and $\sigma_{\ln K_0} = 0.357$. For the material fatigue strength index K_0, the stress unit is ksi. Determine the diameter of the shaft with a reliability 0.99 when its dimension tolerance is $\pm 0.005''$.

Table 3.75: The cyclic combined loading spectrum for Example 3.43

Loading Level #	Number of Cycles n_{Li}	Torque T_i (klb.in)	Bending Moment M_i (klb.in)
1	6,500	3.52	5.75
2	450,000	3.52	3.15

3.44. Use the modified Monte Carlo method to do Problem 3.43.

3.45. The critical section of a rotating shaft is at the shoulder section, as shown in Figure 3.20 and is subjected to cyclic combined bending-torque loading. According to the design specification, the torque T is a constant 4.0 (klb.in). The bending moment M follows a uniform distribution between 2.85 (klb.in) and 3.5 (klb.in). The number of cycles of this combined loading is $n_L = 520,000$ (cycles). The ultimate material strength S_u is 75 (ksi). The three distribution parameters of material fatigue strength index K_0 for the standard specimen under fully reversed bending stress are $m = 8.21$, $\mu_{\ln K_0} = 41.738$, and $\sigma_{\ln K_0} = 0.357$. For the material fatigue strength index K_0, the stress unit is ksi. Determine the diameter of the shaft with a reliability 0.99 when its dimension tolerance is $\pm 0.005''$.

3.46. Use the Monte Carlo method to do Problem 3.45.

APPENDIX A

Computational Methods for the Reliability of a Component

A.1 THE HASOFER–LIND (H-L) METHOD

When all variables are statistically independent normally distributed random variables, the Hasofer–Lind (H-L) method [1, 2] can be used to calculate the reliability of a component with a nonlinear limit state function. The H-L method will linearize the non-limit state function at the design point. The design point is a point on the surface of the limit state function: $g(X_1, X_2, \ldots, X_n) = 0$. Since the design point is generally not known in advance, the H-L method is an iterative process to calculate the reliability of a component with a convergence condition.

Consider the following general nonlinear limit state function, which consists of mutually independently normally distributed random variables:

$$g(X_1, X_2, \ldots, X_n) = \begin{cases} > 0 & \text{Safe} \\ = 0 & \text{Limit state} \\ < 0 & \text{Failure,} \end{cases} \tag{A.1}$$

where X_i $(i = 1, 2, \ldots, n)$ is a normal distributed random variable with corresponding a mean μ_{X_i} and a standard deviation σ_{X_i}. The following equation defines the surface of a limit state function:

$$g(X_1, X_2, \ldots, X_n) = 0. \tag{A.2}$$

The general procedure for the H-L method is explained and displayed here.

Step 1: Pick an initial design point $P^{*0}\left(X_1^{*0}, X_2^{*0}, \ldots, X_n^{*0}\right)$.

The initial design point must be on the surface of the limit state function as specified by Equation (A.2). We can use the mean values for the first $n - 1$ variables, as shown in Equation (A.3):

$$X_i^{*0} = \mu_{X_i} \qquad i = 1, 2, \ldots, n - 1. \tag{A.3}$$

When the actual limit state function is provided, we express X_n^{*0} by using $X_1^{*0}, X_2^{*0}, \ldots$, and X_{n-1}^{*0}. Therefore, the X_n^{*0} can use the following equation to be calculated:

$$X_n^{*0} = g_1\left(X_1^{*0}, X_2^{*0}, \ldots, X_{n-1}^{*0}\right). \tag{A.4}$$

Step 2: Set $\beta = 0$.

This setting is only for the MATLAB program. This setting will make sure that there are at least two iterative loops for the iterative process.

Step 3: Calculate the initial design point in the standard normal distribution space.

We convert a normal distribution X_i into a standard normal distribution Z_i through the following conversion equation:

$$Z_i = \frac{X_i - \mu_{X_i}}{\sigma_{X_i}} \qquad i = 1, \ldots, n. \tag{A.5}$$

The initial design point $P^{*0}\left(X_1^{*0}, X_2^{*0}, \ldots, X_n^{*0}\right)$ in the original normal distributional space can be expressed by $P^{*0}\left(Z_1^{*0}, Z_2^{*0}, \ldots, Z_n^{*0}\right)$ in the standard normal distribution space through Equation (A.5). Z_i^{*0} $(i = 1, \ldots, n)$ can be calculated per Equation (A.6).

$$Z_i^{*0} = \frac{X_i^{*0} - \mu_{X_i}}{\sigma_{X_i}} \qquad i = 1, \ldots, n. \tag{A.6}$$

Step 4: Calculate the reliability index β^{*0} at the design point $P^{*0}\left(Z_1^{*0}, Z_2^{*0}, \ldots, Z_n^{*0}\right)$.

In the H-L method, the limit state function $g\left(Z_1, Z_2, \ldots, Z_n\right)$ is linearized at the initial design point $P^{*0}\left(Z_1^{*0}, Z_2^{*0}, \ldots, Z_n^{*0}\right)$ through the Taylor Series. The Taylor Series coefficient, in this case, will be:

$$G_i\big|_{P*0} = \frac{\partial g\left(Z_1, Z_2, \ldots, Z_n\right)}{\partial Z_i}\bigg|_{at\ P^{*0}\left(Z_1^{*0}, Z_2^{*0}, \ldots, Z_n^{*0}\right)} \qquad i = 1, 2, \ldots, n, \tag{A.7}$$

where $G_i\big|_{P*0}$ means the Taylor Series coefficient for the variable Z_i at the design point $P^{*0}\left(Z_1^{*0}, Z_2^{*0}, \ldots, Z_n^{*0}\right)$. According to the conversion Equation (A.5), we have:

$$\frac{\partial X_i}{\partial Z_i} = \sigma_{X_i}. \tag{A.8}$$

Equation (A.7) can be rewritten as:

$$G_i\big|_{P*0} = \sigma_{X_i} \frac{\partial g\left(X_1, X_2, \ldots, X_n\right)}{\partial X_i}\bigg|_{at\ P^{*0}\left(X_1^{*0}, X_2^{*0}, \ldots, X_n^{*0}\right)}. \tag{A.9}$$

The reliability index β^{*0} per Equations (A.6) and (A.9) will be:

$$\beta^{*0} = \frac{\sum_{i=1}^{n}\left(-Z_i^{*0} G_i\big|_{P*0}\right)}{\sqrt{\sum_{i=1}^{n}\left(G_i\big|_{P*0}\right)^2}}. \tag{A.10}$$

Step 5: Determine the new design point $P^{*1}\left(Z_1^{*1}, Z_2^{*1}, \ldots, Z_n^{*1}\right)$.

The recurrence equation for the iterative process in the H-L method is the following equation.

$$Z_i^{*1} = \frac{-G_i|_{P*0}}{\sqrt{\sum_{i=1}^n \left(G_i|_{P*0}\right)^2}} \beta^{*0} \qquad i = 1, 2, \ldots, n-1. \tag{A.11}$$

Since the new design point $P^{*1}\left(Z_1^{*1}, Z_2^{*1}, \ldots, Z_n^{*1}\right)$ is on the surface of the limit state function $g\left(Z_1^{*1}, Z_2^{*1}, \ldots, Z_n^{*1}\right) = 0$, the Z_n^{*1} will be obtained from the surface of the limit state function. Since we typically still use the limit state function $g\left(X_1, X_2, \ldots, X_n\right) = 0$ to conduct the calculation, we will use the following equations to get the Z_n^{*1}.

We can use the conversion Equation (A.6) to get the first $n-1$ values of the new design point $P^{*1}\left(X_1^{*1}, X_2^{*1}, \ldots X_{n-1}^{*1}, X_n^{*1}\right)$ per Equation (A.12):

$$X_i^{*1} = \mu_{X_i} + \sigma_{X_i} \times Z_i^{*1}. \tag{A.12}$$

The value X_n^{*1} is obtained per Equation (A.4), that is,

$$X_n^{*1} = g_1\left(X_1^{*1}, X_2^{*1}, \ldots, X_{n-1}^{*1}\right). \tag{A.13}$$

When the X_n^{*1} is obtained per Equation (A.13), Z_n^{*1} can be calculated through the conversion Equation (A.14):

$$Z_n^{*1} = \frac{X_n^{*0} - \mu_{X_n}}{\sigma_{X_n}}. \tag{A.14}$$

Now we have the new design point $P^{*1}\left(X_1^{*1}, X_2^{*1}, \ldots, X_n^{*1}\right)$ in the original normal distributional space and the same design point $P^{*1}\left(Z_1^{*1}, Z_2^{*1}, \ldots, Z_n^{*1}\right)$ in the standard normal distributional space.

Step 6: Check convergence condition.

The convergence equation for this iterative process will be the difference $|\Delta\beta^*|$ between the current reliability index and the previous reliability index. Since β is a reliability index, the following convergence condition will provide an accurate estimation of the reliability.

$$|\Delta\beta^*| \leq 0.0001. \tag{A.15}$$

If the convergence condition is satisfied, the reliability of the component will be:

$$R = P\left[g\left(X_1, X_2, \ldots, X_n\right) > 0\right] = \Phi\left(\beta^{*0}\right). \tag{A.16}$$

If the convergence condition is not satisfied, we use this new design point $P^{*1}\left(Z_1^{*1}, Z_2^{*1}, \ldots, Z_n^{*1}\right)$ to replace the previous design point $P^{*0}\left(Z_1^{*0}, Z_2^{*0}, \ldots, Z_n^{*0}\right)$, that is,

$$X_i^{*0} = X_1^{*1}$$
$$Z_i^{*0} = Z_i^{*0} \qquad i = 1, \dots, n \tag{A.17}$$
$$\beta = \beta^{*0}.$$

Then go to Step 4 for a new iterative process again until the convergence condition is satisfied.

Since the H-L method is an iterative process, we should use the program for calculation. The program flowchart for the H-L method is shown in Figure A.1.

A.2   THE RACKWITZ AND FIESSLER (R-F) METHOD

When a limit state function of a component contains at least one non-normal distributed random variables such as log-normal distribution or Weibull distribution, we need to use the R-F (Rackwitz and Fiessler) method [2–4] to calculate the reliability of a component. The R-F method is a modified H-L method. In the R-F method, any non-normally distributed random variable at the design point will be first converted into an equivalent normally distributed random variable. And then the H-L method is applied for calculating the reliability index. Two conditions for calculating the equivalent mean and the equivalent standard deviation of the equivalent normal distribution at the design point are: (1) the PDF of a non-normal distribution variable at the design point will be equal to the PDF of its equivalent normal distribution at the design point; and (2) the CDF of a non-normal distribution variable at the design point will be equal to the CDF of its equivalent normal distribution at the design point.

The following is the general procedure for the R-F method.

Step 1: Calculate the mean for non-normal distributed random variables.

For a clear description of the R-F method procedure, we can rearrange the limit state function per Equation (A.18). In Equation (A.19), the first r random variables are non-normally distributed random variables, and the rest $(n - r)$ random variables are normally distributed random variables.

$$g(X_1, \dots, X_r, X_{r+1}, \dots, X_n) = \begin{cases} > 0 & \text{Safe} \\ = 0 & \text{Limit state} \\ < 0 & \text{Failure.} \end{cases} \tag{A.18}$$

The surface of this limit state function is

$$g(X_1, \dots, X_r, X_{r+1}, \dots, X_n) = 0. \tag{A.19}$$

For non-normally distributed random variable, we can use their PDFs to calculate their means: μ_{X_i} $(i = 1, 2, \dots, r)$.

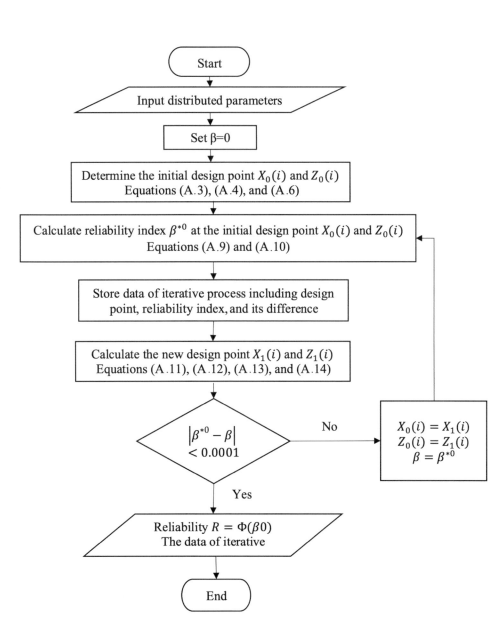

Figure A.1: The program flowchart for the H-L method.

Step 2: Pick an initial design point $P^{*0}\left(X_1^{*0}, X_2^{*0}, \ldots, X_n^{*0}\right)$.

The initial design point could be any point. But it must be on the surface of the limit state function as specified by Equation (A.19). We can use the mean values for the first $n - 1$ variables, as shown in Equation (A.20):

$$X_i^{*0} = \mu_{X_i} \qquad i = 1, 2, \ldots, n - 1. \tag{A.20}$$

X_n^{*0} can be determined through the surface of the limit state function of Equation (A.19). When the actual limit state function is provided, we can rearrange express X_n^{*0} by using $X_1^{*0}, X_2^{*0}, \ldots,$ and X_{n-1}^{*0} as shown in Equation (A.21):

$$X_n^{*0} = g_1\left(X_1^{*0}, X_2^{*0}, \ldots, X_{n-1}^{*0}\right). \tag{A.21}$$

Now, we have the initial design point $P^{*0}\left(X_1^{*0}, X_2^{*0}, \ldots X_{n-1}^{*0}, X_n^{*0}\right)$.

Step 3: Set $\beta = 0$.

This setting is only for the MATLAB program. This setting will make sure that the iterative process will have at least two iterative loops.

Step 4: The mean and standard deviation at the design point $P^{*0}\left(X_1^{*0}, X_2^{*0}, \ldots X_{n-1}^{*0}, X_n^{*0}\right)$.

For non-normally distributed random variables, we convert them into equivalent normal distributed random variables and calculate its equivalent mean and equivalent standard deviation per Equation (A.22):

$$
\begin{aligned}
z_{X_i}^{*0} &= \Phi^{-1}\left[F_{X_i}\left(X_i^{*0}\right)\right] = norminv\left(F_{X_i}\left(X_i^{*0}\right)\right) \\
\sigma_{X_i eq} &= \frac{1}{f_{X_i}\left(X_i^{*0}\right)} \phi\left(z_{X_i}^{*0}\right) \qquad i = 1, 2, \ldots, r \\
\mu_{X_i eq} &= x_i^{*0} - z_{X_i}^{*0} \times \sigma_{X_i eq},
\end{aligned}
\tag{A.22}
$$

where x_i^{*0} is the value of the non-normally distributed random variable X_i at the design point $P^{*0}\left(X_1^{*0}, X_2^{*0}, \ldots X_{n-1}^{*0}, X_n^{*0}\right)$. $f_{X_i}\left(x_i^{*0}\right)$ and $F_{X_i}\left(x_i^{*0}\right)$ are the PDF and the CDF of the non-normally distributed random variable X_i at the design point X_i^{*0}. $\mu_{X_i eq}$ and $\sigma_{X_i eq}$ are the equivalent mean and the equivalent standard deviation of the equivalent normally distributed random variable at the design point x_i^{*0}.

Now every random variable in the limit state function in Equation (A.18) at the design point P^{*0} are normally distributed random variables. The mean and standard deviation of these normally distributed random variables are

$$\mu_{X_i} = \begin{cases} \mu_{X_{ieq}} & i = 1, 2, \ldots, r \\ \mu_{X_i} & i = r + 1, \ldots n \end{cases} \tag{A.23}$$

$$\sigma_{X_i} = \begin{cases} \sigma_{X_{ieq}} & i = 1, 2, \ldots, r \\ \sigma_{X_i} & i = r + 1, \ldots n. \end{cases} \tag{A.24}$$

Step 5: Calculate the initial design point P^{*0} in the standard normal distribution space.

In the standard normal distribution space, the initial design point $P^{*0}\left(X_1^{*0}, \ldots, X_n^{*0}\right)$ can be expressed as $P^{*0}\left(Z_1^{*0}, Z_2^{*0}, \ldots, Z_n^{*0}\right)$ in the standard normal distribution space through Equation (A.25):

$$Z_i^{*0} = \frac{X_i^{*0} - \mu_{X_i}}{\sigma_{X_i}} \qquad i = 1, \ldots, n. \tag{A.25}$$

Step 6: Calculate the reliability index β^{*0} at the design point $P^{*0}\left(Z_1^{*0}, Z_2^{*0}, \ldots, Z_n^{*0}\right)$.

Per Equation (A.26), we can calculate the Taylor series coefficients, and per Equation (A.27) we can calculate the reliability index β^{*0}:

$$G_i|_{P*0} = \sigma_{X_i} \left. \frac{\partial g\left(X_1, X_2, \ldots, X_n\right)}{\partial X_i}\right|_{at\ P^{*0}\left(X_1^{*0}, X_2^{*0}, \ldots, X_n^{*0}\right)} \qquad i = 1, 2, \ldots, n \tag{A.26}$$

$$\beta^{*0} = \frac{\sum_{i=1}^{n}\left(-Z_i^{*0} G_i|_{P*0}\right)}{\sqrt{\sum_{i=1}^{n}\left(G_i|_{P*0}\right)^2}}. \tag{A.27}$$

Step 7: Determine the new design point $P^{*1}\left(Z_1^{*1}, Z_2^{*1}, \ldots, Z_n^{*1}\right)$ for the iterative process.

The recurrence equations for the iterative process are the following equations:

$$Z_i^{*1} = \frac{-G_i|_{P*0}}{\sqrt{\sum_{i=1}^{n}\left(G_i|_{P*0}\right)^2}}\beta^{*0} \qquad i = 1, 2, \ldots, n - 1. \tag{A.28}$$

Since the new design point $P^{*1}\left(Z_1^{*1}, Z_2^{*1}, \ldots, Z_n^{*1}\right)$ is on the surface of the limit state function $g\left(Z_1^{*1}, Z_2^{*1}, \ldots, Z_n^{*1}\right) = 0$, the Z_n^{*1} will be obtained from the surface of the limit state function. Since we typically still use the limit state function $g\left(X_1, X_2, \ldots, X_n\right) = 0$ to conduct the calculation, we will use the following equations to get the Z_n^{*1}.

We can use the conversion Equation (A.25) to get the first $n - 1$ values of the new design point $P^{*1}\left(X_1^{*1}, X_2^{*1}, \ldots X_{n-1}^{*1}, X_n^{*1}\right)$, that is:

$$X_{1i}^{*1} = \mu_{X_i} + \sigma_{X_i} \times Z_i^{*1}. \tag{A.29}$$

Per the surface of the limit state function Equation (A.19), we express X_n^{*1} as the function of $X_1^{*1}, X_2^{*1}, \ldots,$ and X_{n-1}^{*1}, as shown in Equation (A.30):

$$X_n^{*1} = g_1\left(X_1^{*1}, X_2^{*1}, \ldots, X_{n-1}^{*1}\right). \tag{A.30}$$

After the X_n^{*1} is obtained from Equation (A.30), Z_n^{*1} can be calculated through the conversion Equation (A.31):

$$Z_n^{*1} = \frac{X_n^{*0} - \mu_{X_n}}{\sigma_{X_n}}. \tag{A.31}$$

Now we have the new design point $P^{*1}\left(X_1^{*1}, X_2^{*1}, \ldots, X_n^{*1}\right)$ in original normal distribution space and the same design point $P^{*1}\left(Z_1^{*1}, Z_2^{*1}, \ldots, Z_n^{*1}\right)$ in the standard normal distribution space.

Step 8: Check convergence condition.

The convergence equation for this iterative process will be the difference $\left|\Delta\beta^{*0}\right|$ between the current reliability index and the previous reliability index. Since β is a reliability index, the following convergence condition will provide an accurate estimation of the reliability:

$$\left|\Delta\beta^{*0}\right| \leq 0.0001. \tag{A.32}$$

If the convergence condition is satisfied, the reliability of the component will be:

$$R = P\left[g\left(X_1, X_2, \ldots, X_n\right) > 0\right] = \Phi\left(\beta^{*0}\right). \tag{A.33}$$

If the convergence condition is not satisfied, we use this new design point $P^{*1}\left(X_1^{*1}, X_2^{*1}, \ldots, X_n^{*1}\right)$ to replace the previous design point $P^{*0}\left(X_1^{*0}, X_2^{*0}, \ldots, X_n^{*0}\right)$, that is,

$$\begin{aligned} X_i^{*0} &= X_i^{*1} \qquad i = 1, \ldots, n \\ \beta &= \beta^{*0}. \end{aligned} \tag{A.34}$$

Then, we go to Step 4 for a new iterative process again until the convergence condition is satisfied.

The program flowchart for the R-F method is shown in Figure A.2.

A.3   THE MONTE CARLO METHOD

A general limit state function $g\left(X_1, X_2, \ldots, X_n\right)$ of a component is the function of random variables $X_1, X_2, \ldots,$ and X_n. Therefore, it is also a random variable. Per the definition of probability, we can use the relative frequency to estimate the reliability when the number of sample data of the limit state function is sufficiently big [2]. The Monte Carlo method [2, 3, 5] relies on repeated random sampling to obtain the numerical value of the limit state function $g\left(X_1, X_2, \ldots, X_n\right)$ for estimating the relative frequency.

Basic concepts and procedure for the Monte Carlo method are as follows.

Step 1: Uniformly and randomly generate one sample value for each random variable per its corresponding probabilistic distribution. Let x_i^{*j} $(i = 1, 2, \ldots, n)$ to be the sample data in the

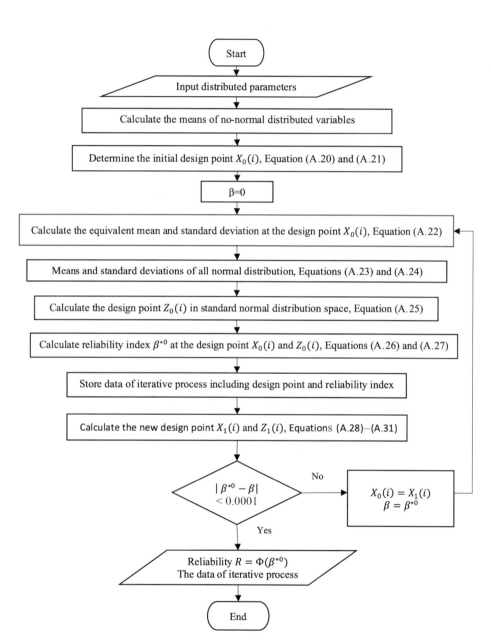

Figure A.2: The program flowchart for the R-F method.

jth trial of the virtual experiment. Here, the subscript i in x_i^{*j} refers to the ith random variable X_i. The superscript j in x_i^{*j} refers to the jth trial. The x_i^{*j} is the sample value of the random variable X_i in the jth trial of the virtual experiment.

Step 2: Use x_i^{*j} ($i = 1, 2, \ldots, n$) in the limit state function to get a trial value of the limit state function. Per the definition of the limit state function, when the trial value: $g\left(x_1^{*j}, x_2^{*j}, \ldots, x_n^{*j}\right)$ of the limit state function of the component is larger than or equal to zero, the component is safe. When the trail value: $g\left(x_1^{*j}, x_2^{*j}, \ldots, x_n^{*j}\right)$ of the limit state function of the component is less than zero, the component is a failure. We can use VT^{*j} to represent the trial result:

$$VT^{*j} = \begin{cases} 1 & \text{when } g\left(x_1^{*j}, x_2^{*j} \ldots, x_n^{*j}\right) \geq 0 \\ 0 & \text{when } g\left(x_1^{*j}, x_2^{*j}, \ldots, x_n^{*j}\right) < 0, \end{cases} \tag{A.35}$$

where VT^{*j} is the trial result of the jth trial of the virtual experiment. The value "1" of the VT^{*j} indicates a safe status of the component. The value "0" of the VT^{*j} indicates a failure status of the component.

Step 3: Repeat Steps 1 and 2 until enough number of trials N have been conducted.

Since the limit state function of a mechanical component is typically not too complicated, we can use $N = 15{,}998{,}400$, which is big enough for a critical component with a reliability 0.9999.

Step 4: The relative frequency of the component with a safe status in total trial N will be the probability of the event $g(X_1, X_2, \ldots, X_n) \geq 0$. Therefore, the reliability of the component will be

$$R = P[g(X_1, X_2, \ldots, X_n) \geq 0] = \frac{\sum_{j=1}^{N} \left(VT^{*j}\right)}{N}. \tag{A.36}$$

The probability of the component failure F will be:

$$F = 1 - R = 1 - \frac{\sum_{j=1}^{N} \left(VT^{*j}\right)}{N}. \tag{A.37}$$

Step 5: Calculate the relative errors.

In the Monte Carlo method, the relative error between the true value of the probability of the component failure and the estimated value in Equation (A.37) will become smaller when the trail number N increases. For a 95% confidence level, the relationship [2, 4] between the relative error ε and the trial number N is:

$$\varepsilon = 2\sqrt{\frac{1 - F}{N \times F}}, \tag{A.38}$$

where ε is the relative error of the probability of component failure with a 95% confidence level.

The Monte Carlo method needs to have a huge amount of trials for an acceptable accuracy of the estimated reliability. MATLAB software has a function to generate a matrix $1 \times N$ of random numbers of a specified distributed random variable, as shown in Equation (A.39):

$$RX_i = random \; ('name', \; A, B, \; 1, \; N),$$ (A.39)

where RX_i is a matrix $1 \times N$ with N of random samplings of a distributed random variable X_i. *random* is the MATLAB command for generating a random number. *'name'* is to specify the type of distribution. A and B are the distribution parameters of the distributed random variables. "1, N" in Equation (A.39) means that the matrix for these random number will be stored as one row with N column.

The flowchart of a MATLAB program for the Monte Carlo method is displayed in Figure A.3.

A.4   REFERENCES

[1] Hasofer, A. M. and Lind, N., An exact and invariant first-order reliability format, *Journal of Engineering Mechanics, ASCE*, vol. 100, no. EM1, pp. 111–121, February 1074. 213

[2] Le, Xiaobin, *Reliability-Based Mechanical Design, Volume 1: Component under Static Load*, Morgan & Claypool Publishers, San Rafael, CA, 2020. 213, 216, 220, 222

[3] Andrzej, S. N. and Collins, K. R., *Reliability of Structures*, 2nd ed., CRC Press, Boca Raton, FL, 2013. 220

[4] Rackwitz, R. and Fiessler, B., Structural reliability under combined random load sequences, *Computers and Structures*, vol. 9, pp. 489–494, 1978. DOI: 10.1016/0045-7949(78)90046-9. 216, 222

[5] Singiresu, S. R., *Reliability Engineering*, Pearson, 2015. 220

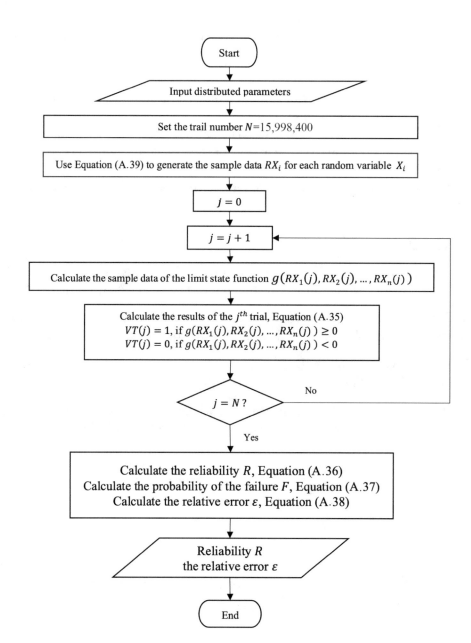

Figure A.3: The flowchart of a MATLAB program for the Monte Carlo method.

APPENDIX B

Samples of MATLAB® Programs

B.1 THE H-L METHOD FOR EXAMPLE 2.6

```
% The H-L method for Example 2.6
% The Limit State function: g(ka,kc,se,d,fa)
% Input the distribution parameters mx-mean,
% sz-standard deviation
clear,       % Clear the memory
mx=[0.905,0.774,24.7,1.25,8.5];       % The means
sx=[0.0543,0.1262,2.14,0.0125,1.2];   % The standard
                                      % deviations
n=5;      % Total four random variables
beta=0;  % Set beta=0
% Pick an initial design point x0(i)
for i=1:n-1
   x0(i)=mx(i); %Use the means for the first n-1 variable
end
% Use surface of the limit state function to calclaue the
% value for the last random variable
x0(n)=x0(1)*x0(2)*x0(3)*(61.5*(x0(4))^2-15.279)/78.290;
% Initial point in standard normal distribution space
for i=1:n
   z0(i)=(x0(i)-mx(i))/sx(i);
end
% Start iterative process
for j=1:1000
% Calculate the reliability beat0
% The Tylor series coefficent
G0(1)=sx(1)*x0(2)*x0(3);
G0(2)=sx(2)*x0(1)*x0(3);
G0(3)=sx(3)*x0(1)*x0(2);
```

```matlab
G0(4)=sx(4)*78.290*x0(5)*61.5*2*x0(4)/(61.5*(x0(4))^2-15.279)^2;
G0(5)=sx(5)*(-78.290)/(61.5*(x0(4))^2-15.279);
g00=0;
z00=0;
for i=1:n
g00=g00+G0(i)^2;
z00=z00+(-1)*z0(i)*G0(i);
end
Gi0=g00^0.5;
% Calculate the reliability index beta0
beta0=z00/Gi0;
% Data of iterative process
  for i=1:n
      ddp(j,i)=x0(i);
  end
ddp(j,n+1)=beta0;
ddp(j,n+2)=abs(beta0-beta);
% New design proint
% The values for the first n-1 random variable
  for i=1:n-1
      z1(i)=(-1)*beta0*G0(i)/Gi0;
      x1(i)=sx(i)*z1(i)+mx(i);
  end
% The value for the last random variable from the
% surface of the limit state function
x1(n)=x1(1)*x1(2)*x1(3)*(61.5*(x1(4))^2-15.279)/78.290;
z1(n)=(x1(n)-mx(n))/sx(n);
% Check the convengence condition
  if ddp(j,n+2)<=0.0001;
  break
  end
% Use new design point to replace previous design point
  for i=1:n
      z0(i)=z1(i);
      x0(i)=x1(i);
  end
  beta=beta0;
 end
% Calculate and display reliability
```

```
format short e
disp('reliability')
R=normcdf(beta0)
% Displaye iterative process and write it to Excel file
disp(ddp)
xlswrite('example2_6',ddp); % Write the iterative process
                            % into a Excel file
```

B.2   THE R-F METHOD FOR EXAMPLE 2.7

```
% The R-F method for Example 2.7
% The Limit State function: g(M,ka,se, kf, d)
clear; % Clean memory
% Input the distribution parameters dp1 and dp2
% Means or the first distribution parameter
dp1=[0.315,0.905,24.7,1.562,1.250];
% Standard deviation or the second distribution parameter
dp2=[0.142,0.0543,2.14,0.1250,0.00125];
kb=0.8507; % Size mofification factor
r=1;       %Number of non-normal distribution
n=5;       % Total five random variables
% Calculate the mean of lognorm M and the initial
% point x0(i)
x0(1)=exp(dp1(1)+dp2(1)^2/2);
for i=2:n-1
   x0(i)=dp1(i);
end
% Use the surface of the limit state function
% to calculate x0(n)
x0(5)=(x0(4)*32*x0(1)/pi/(x0(2)*kb*x0(3)))^(1/3);
beta=0; %Set the beta =0
% Iterative process starting
for j=1:1000
%Calculate the equivalent mean and standard deviation
zteq=norminv(logncdf(x0(1),dp1(1),dp2(1)));
steq=normpdf(zteq)/lognpdf(x0(1),dp1(1),dp2(1));
mteq=x0(1)-zteq*steq;
% Mean and standard deviation matrix
meq(1)=mteq;
```

```
    seq(1)=steq;
    for i=2:n
       meq(i)=dp1(i);
       seq(i)=dp2(i);
    end
    % Calculate z0(i) in standrad normal distribution space
    for i=1:n
       z0(i)=(x0(i)-meq(i))/seq(i);
    end
    % Calculate the Taylor Series Coefficient
    Gi(1)=seq(1)*(-x0(4))*32/pi()/(x0(5))^3;
    Gi(2)=seq(2)*kb*x0(3);
    Gi(3)=seq(3)*x0(2)*kb;
    Gi(4)=seq(4)*(-32)*x0(1)/pi()/(x0(5))^3;
    Gi(5)=seq(5)*x0(4)*96*x0(1)/pi()/(x0(5))^4;
    g00=0;
    z00=0;
    for i=1:n
    g00=g00+Gi(i)^2;
    z00=z00+(-1)*z0(i)*Gi(i);
    end
    Gi0=g00^0.5;
    %Calculate the reliability index beta0
    beta0=z00/Gi0;
    % Data of iterative process
    for i=1:n
       ddp(j,i)=x0(i);
    end
    ddp(j,n+1)=beta0;
    ddp(j,n+2)=abs(beta0-beta);
    % New design proint
      for i=1:n-1
         z1(i)=(-1)*beta0*Gi(i)/Gi0;
         x1(i)=seq(i)*z1(i)+meq(i);
      end
    % Use the surface of the limit state function to
    % calculate x1(n)
      x1(5)=(x1(4)*32*x1(1)/pi/(x1(2)*kb*x1(3)))^(1/3);
      z1(5)=(x1(5)-meq(5))/seq(5);
```

```
% Check the convengence condition
  if ddp(j,n+2)<=0.0001;
  break
  end
% Use new design point to replace previous design point
  for i=1:5
      x0(i)=x1(i);
  end
  beta=beta0;
end
% Calculate and display reliability
format short e
disp('reliability')
R=normcdf(beta0)
% Display e iterative process and write it to Excel file
disp(ddp)
xlswrite('example2.7',ddp); % Write the data in a
                            % Excel file
```

B.3   THE MONTE CARLO METHOD FOR EXAMPLE 2.8

```
% The Monte Carlo method for Example 2.8
% The Limit State function: g(M,ka, se, kf, d)
clear; % clean memory
%input the distribution parameters dp1 and dp2
dp1=[0.772,24.7,2.0,2.0,11.5];        %The means
dp2=[0.0757,2.14,0.0025,0.0025,1.5]; %The standard deviation
kb=0.8507;
% The trail number
N=15998400
% Generate random numbers for each random variable
Rka=random('norm',dp1(1),dp2(1),1,N);
Rse=random('norm',dp1(2),dp2(2),1,N);
Rh=random('norm',dp1(3),dp2(3),1,N);
Rb=random('norm',dp1(4),dp2(4),1,N);
RMa=random('norm',dp1(5),dp2(5),1,N);
nn=0;
for k=1:N
   % Calculate the value of the limit state function
```

```
      tr=0.826*Rka(k)*Rse(k)-369*RMa(k)/(61.5*Rb(k)*Rh(k)^2-123);
      if tr>0
      % Number of the safe status
         nn=nn+1;
      end
end
% Reliability of the component
R=nn/N
% The failure probability of the component
F=1-R
% The percent relative error of the failure probability
rerror=2*(R/N/F)^0.5
% The range of the error for the failure probability
% and the reliability
erange=F*rerror
```

B.4   THE M-H-L METHOD FOR EXAMPLE 3.3

```
% The modified H-L method for Example 3.3
% The Limit State function:
% g(Sy,F,Kt, d)=Sy-Kt*4*F/(pi*d^2)
clear;     %Clear the momery
% Input the distribution parameters m-first or mean,
% s-second or standard deviation
mx=[32.2,28.72];
sx=[3.63,2.87];
% Prelimary design for Kt
mx(3)=1.9;       % Preliminary Kt
sx(3)=1.9*0.05; % The stadard deviation of Kt
D=3.25;        % Bigger diameter in the stress
               % concentration area
r=0.125;       % Fillet radius
sd=0.00125; % Standard deviation of dimension d
R=0.99;            % The required reliability
beta=norminv(R); % Reliability index
% The initial design point x0(i), i=1,...,3
for i=1:3
    x0(i)=mx(i);
end
```

```
% Use the limit state function to determine x0(4)
x0(4)=(4*x0(3)*x0(2)/pi/x0(1))^0.5;
% Store initial design point
for i=1:4
    dpp(1,i)=x0(i);
end
% Iterative process
for j=2:1000
% The Tylor series coefficent
G1=sx(1)*1;
G2=sx(2)*(-4)*x0(3)/pi/x0(4)^2;
G3=sx(3)*(-4)*x0(2)/pi/x0(4)^2;
Gd=sd*8*x0(3)*x0(2)/pi/x0(4)^3;
G0=(G1^2+G2^2+G3^2+Gd^2)^0.5;
%Calculate the new design point
x1(1)=mx(1)+sx(1)*beta*(-G1)/G0;
x1(2)=mx(2)+sx(2)*beta*(-G2)/G0;
x1(3)=mx(3)+sx(3)*beta*(-G3)/G0;
% Use the limit state function to determine x1(4)
x1(4)=(4*x1(3)*x1(2)/pi/x1(1))^0.5;
% Update the dimension-dependent Kt
dd=x1(4)-sd*beta*(-Gd)/G0;      % New value for
                                % the dimension
mx(3)=StressAxial( D, dd, r ); % Update Kt
sx(3)=0.05*mx(3);
% Data of iterative process
  for i=1:4
     dpp(j,i)=x1(i);
  end
  dpp(j,4+1)=abs(dpp(j,4)-dpp(j-1,4));
  % Check the convengence condition
  if dpp(j,4+1)<=0.0001;
  break
  end
  % Use new design point to replace previous
  % design point
  for i=1:4
    x0(i)=x1(i);
  end
```

```
  end
% Display reliability
format short e
% Displaye iterative process and write it to
% Excel file
disp(dpp)
xlswrite('example6.3',dpp)
%Display the mean of the dimension
dmean=x1(4)-sd*beta*(-Gd/G0)
% the end of the program
% The stress concentration Kt for a stepped shaft under
% axial loading.
function [ kt ] = StressAxial( D, d, r )
Dt = [2.0 1.5 1.3 1.2 1.15 1.1 1.07 1.05 1.02 1.01]';
At = [1.01470 .99957 .99682 .96272 .98084 .98450 .98498
        1.00480 1.01220 .98413]';
bt = [-.30035 -.28221 -.25751 -.25527 -.24485 -.20818
        -.19548 -.17076 -.12474 -.10474]';
% Compute the diameter ratio and then interpolate A and
% b from the tables
DD = D / d;
A = interp1 (Dt, At, DD);
b = interp1 (Dt, bt, DD);
% Compute the stress concentration factor
kt = A * (r / d) ^ b;
end
```

B.5   THE M-R-F PROGRAM FOR EXAMPLE 3.5

```
% The Modified R-F method for Example 3.5
% The Limit State function:
% g(T, Ssy, Kts, d)=Ssy-Kts*16T/(pi*d^3)
clear
% Input the distribution parameters
mx=[20,31]; % Mean or the first parameter
sx=[3,2.4,]; % Stadard deviatio or the second parameter
% Prelimary design for Kts
mx(3)=1.6;       % Preliminary Kts
sx(3)=1.6*0.05;  % The stadard deviation of Kts
```

```
sd=0.00125;        % The standard deviation of dimension
R=0.99;            % The required reliability
beta=normcdf(R),   % Reliability index beta
D=3.25;            % The bigger radius in stress
                   % concentration area
r=1/16;            % The fillet radius
% Calculate the mean of T (Weibull) and the initial
% point x0(i)
% Initial design point
x0(1)=mx(1)*gamma(1/sx(1)+1);
x0(2)=mx(2);
x0(3)=mx(3);
% Use the limit state function to determine x0(4)
x0(4)=(16*x0(1)*x0(3)/pi/x0(2))^(1/3);
% Store initial design point
for i=1:4
    ddp(1,i)=x0(i);
end
% Iterative process starting
for j=2:1000
% Calculate the equivalent mean and standard deviation
zteq=norminv(wblcdf(x0(1),mx(1),sx(1)));
steq=normpdf(zteq)/wblpdf(x0(1),mx(1),sx(1));
mteq=x0(1)-zteq*steq;
% Mean and standard deviation matrix at the
% design point
meq(1)=mteq;
seq(1)=steq;
    for i=2:3
        meq(i)=mx(i);
        seq(i)=sx(i);
    end
% Calculate the Taylor Series Coefficient
G1=-seq(1)*16*x0(3)/pi/x0(3)^3;
G2=seq(2);
G3=seq(3)*(-16)*x0(1)/pi/x0(4)^3
Gd=sd*48*x0(1)/pi/x0(3)^4;
G0=(G1^2+G2^2+G3^2+Gd^2)^0.5;
% New design proint
```

```
x1(1)=-seq(1)*beta*G1/G0+meq(1);
x1(2)=-seq(2)*beta*G2/G0+meq(2);
x1(3)=-seq(3)*beta*G3/G0+meq(3);
% Use the limit state function to determine x1(4)
x1(4)=(16*x1(1)*x0(3)/pi/x1(2))^(1/3);
% Update the stress concentration factor
dd=x1(4)-sd*beta*(-Gd)/G0;  % the new value of the
                            % dimension
mx(3)=StressTorsion( 3.25, dd, r );
s(3)=0.05*mx(3);
  %store iterative process
  for i=1:4
  ddp(j,i)=x1(i);
  end
  ddp(j,4+1)=abs(ddp(j-1,4)-ddp(j,4));
% Check the convengence condition
  if ddp(j,4+1)<=0.0001;
  break
  end
% Use new design point to replace previous design point
  for i=1:4
      x0(i)=x1(i);
  end
 end
format short e
% Display e iterative process and write it to Excel file
disp(ddp)
xlswrite('example6.5',ddp)
display('mean of the dimension with given reliability')
md=x1(4)-sd*beta*(Gd/G0)
% This routine computes the stress concentration factor
% for an torsional load
function [ kt ] = StressTorsion( D, d, r )
Dt = [2.0 1.33 1.20 1.09]';
At = [.86331 .84897 .83425 .90337]';
bt = [-.23865 -.23161 -.21649 -.12692]';
% Compute the diameter ratio and then interpolate A and
% b from the tables
DD = D / d;
```

```
if (DD > Dt(1))
    DD = Dt(1);
end
if (DD < Dt(4))
    DD = Dt(4);
end
A = interp1 (Dt, At, DD);
b = interp1 (Dt, bt, DD);
% Compute the stress concentration factor
kt = A * (r / d) ^ b;
end
```

B.6   THE MODIFIED MONTE CARLO METHOD FOR EXAMPLE 3.7

```
%The modified Monte Carlo method for Example 3.7
%Limit state function g(Sy,F,d)=Sy-4F/(pi*d^2)
%Input data
clear
mx=[34.5,7.0];    % The first parameter or mean
sx=[3.12,9.0];    % The second parameter or mean
sd=0.00125;       % The standard deviation of the dimension
R=0.99;           % The required reliability
% The first value for mean of d
xstar=norminv(1-R,mx(1),sx(1));
% Mean for F
mf=(mx(2)+sx(2))/2;
% The initial value of the dimension
mdd=(4*mf/pi/xstar)^0.5
N=15998400;   % the trial number
Rsy=random('norm',mx(1),sx(1),1,N); % Random samples
                                    % for Sy
Rf=random('unif',mx(2),sx(2),1,N);  % Random samples
                                    % for F
for K=1:2000
nn=0;
K
md=mdd+0.001; % Iterative dimension with an
              % incremental 0.001"
```

```
Rd=random('norm',md,sd,1,N);        % Random samples for d
for j=1:N
   fj=Rsy(j)-4*Rf(j)/pi/Rd(j)^2; % Value of the limit
                                 % state function
   if fj>0
      nn=nn+1;
   end
end
Rstar=nn/N; % the reliability of component with d
% Store the iterative process
dpp(K,1)=md;
dpp(K,2)=Rstar;
dpp(K,3)=Rstar-R;
% Check the convergence condition
if dpp(K,3)>0.0001
   break
end
mdd=md;
end
format short e
% Displaye iterative process and write it to Excel file
disp(dpp)
xlswrite('example6.7',dpp)
display('The mean of the dimension with the required reliability')
md
```

Author's Biography

XIAOBIN LE

Xiaobin Le, Ph.D., P.E., received a BS in Mechanical Engineering in 1982 and an MS in Mechanical Engineering in 1987 from Jiangxi University of Science and Technology, Ganzhou, Jiangxi. He received his first Ph.D. in Mechanical Design of Mechanical Engineering from Shanghai Jiao Tong University, Shanghai, in 1993, and his second Ph.D. in Solid Mechanics of Mechanical Engineering from Texas Tech University, Lubbock, Texas, in 2002.

He is currently a professor in the Mechanical Engineering Department at Wentworth Institute of Technology, Boston, Massachusetts. His teaching and research interests are Computer-Aided Design, Mechanical Design, Finite Element Analysis, Fatigue Design, Solid Mechanics, Engineering Reliability, and Engineering Education Research.

Printed in the United States
by Baker & Taylor Publisher Services